How To Thir

DGM: I
with tha
Isaac. I

JFP: I d
inspire n
little pie

How To Think About Cities

Deborah G. Martin
and Joseph Pierce

polity

Copyright © Deborah G. Martin & Joseph Pierce 2023

The right of Deborah G. Martin & Joseph Pierce to be identified as Author of this Work has been asserted in accordance with the UK Copyright, Designs and Patents Act 1988.

First published in 2023 by Polity Press

Polity Press
65 Bridge Street
Cambridge CB2 1UR, UK

Polity Press
111 River Street
Hoboken, NJ 07030, USA

All rights reserved. Except for the quotation of short passages for the purpose of criticism and review, no part of this publication may be reproduced, stored in a retrieval system or transmitted, in any form or by any means, electronic, mechanical, photocopying, recording or otherwise, without the prior permission of the publisher.

ISBN-13: 978-1-5095-3618-4
ISBN-13: 978-1-5095-3619-1(pb)

A catalogue record for this book is available from the British Library.

Library of Congress Control Number: 2022939160

Typeset in 10.5 on 12.5pt Sabon
by Fakenham Prepress Solutions, Fakenham, Norfolk NR21 8NL
Printed and bound in Great Britain by TJ Books Ltd, Padstow, Cornwall

The publisher has used its best endeavours to ensure that the URLs for external websites referred to in this book are correct and active at the time of going to press. However, the publisher has no responsibility for the websites and can make no guarantee that a site will remain live or that the content is or will remain appropriate.

Every effort has been made to trace all copyright holders, but if any have been overlooked the publisher will be pleased to include any necessary credits in any subsequent reprint or edition.

For further information on Polity, visit our website: politybooks.com

Contents

Boxes	vi
Acknowledgments	vii
1 Introduction: Cities are Places	1
2 City of London: A Machine for Living/The Seat of Wealth	25
3 Tehran: Islamic Developmentalism/Diverse Cosmopolitanism	48
4 Worcester: Local Economic Engine/Regional Forest Under Threat	72
5 Portland: Paradise of Environmentalism/Legacy of Exclusionary Racism	97
6 Chongqing: International Cyberpunk Marvel/National Policy Innovator	123
7 Jerusalem: Religious Tourist Destination/Ethno-National Citadel	153
8 Conclusion: The Impossibilities of Fully Knowing a City	176
Notes	192
References	196
Index	221

Boxes

Urban and the City	3
History of Place Theory	7
Conceptual Framing and Place Frames	11
Structure and Agency	15
Capitalism and Urban Political Economy as a Comparison of Systems	29
Urban Economic Processes	33
Modernism and Urban Design	37
Urban Cultural Geographies	43
The Growth Coalition	55
Urban Agglomeration	59
Freedom and Diversity in Cities	65
Municipal Development and Finance Strategies	77
Policy Mobilities	87
Urban Environments	89
Sprawl, Density, and Urban Growth Boundaries	101
Ecomodernism	107
Residential Segregation and Redlining in America	111
Race in Early American Urban Theory: The Chicago School and W. E. B. Du Bois	115
Black Geographies	119
City Branding	129
Asian Futurism, Sinofuturism, and Orientalism	137
Urban Migration and *Hukou* in China	143
Land Markets	149
Toponyms and Demonyms	155
Landscapes and Power	167

Acknowledgments

This book, like all big writing projects, has been supported in ways small and large by a whole host of people, not all of whom we can manage to name here.

DGM: Clark University, and the Graduate School of Geography in particular, has been a supportive, nurturing, and also challenging (mostly in a good way) environment as I wrote this book. Sabbatical support, the students in my classes, and the intellectual environment as a whole provided a robust context for the writing process. Undergraduate students at Clark University, particularly within the Human-Regional Environmental Observatory (HERO) program, have enthusiastically adopted the notion of "relational places," integrating the idea into some of their work. Their adaptability to the idea of multiple place frames helped us to imagine some of our future readers. I am grateful to be at an institution where I am challenged and able to contribute in myriad ways, and at the same time feel valued.

JFP: Most of the writing of this book occurred while I have been situated in the Department of Geography and Environment at the University of Aberdeen, where colleagues have welcomed me with open arms, even in pandemic times. The experience of living in the United Kingdom (and thus in Europe!) has widened my gaze and made the book better. Thank you to the dedicated folks at Aberdeen whose collegiality made this project possible.

TOGETHER, IN CHORUS: We would like to thank Emma Longstaff, a commissioning editor at Polity in 2014, when the seeds of the idea for this book were planted. Jonathan Skerrett and Karina Jákupsdóttir followed up as patient and helpful stewards of the project. Azadeh (Azi) Hadizadeh Esfahani and Amy Yueming Zhang generously provided their insights and substantive engagement as (co-)authors on two of the chapters. Sophia Jacobson and Leslie Gross-Wyrtzen provided useful details about sights and sites in Jerusalem. Along our way, many colleagues, students, and friends (lines here blur) have engaged with and pushed our thinking on cities and place. Kate Boyer, Sarah Elwood, Jody Emel, Katherine Hankins, Helga Leitner, Mary Lawhon, Gordon MacLeod, James Murphy, Mark Purcell, John Rogan, Eric Sheppard, Olivia Williams, and many other interlocutors who have helped make these ideas stronger and clearer: we thank you.

–1–
Introduction: Cities are Places

It is harder than you probably think it is to think clearly about cities.

Are cities defined by clusters of high-rise buildings connected by large roads or highways? Are they rows of restaurants offering a wide variety of ethnic foods on walkable streets where neighbors know and recognize one another? Are they chaotic crowded sidewalks or buses with people jostling and competing with one another for access to doors or seats with windows? Cities are all of these, but they are not only these. Trying to make sense of cities requires some way of organizing these scenes or features: a strategy for making sense of their concatenated elements.

Scholars of cities usually organize these scenes or features into analytical categories like economy, politics, or culture. They often focus on how one of those dimensions shapes or defines being in a city. This book makes the argument that the most productive way to understand a city is to think of it as multiple cities at once. Cities are places where what is meaningful to people is the way they *combine* various processes, ones that occur in domains like economy, politics, and culture. Yet the most visible processes are not always the most relevant ones for analysis, and different urban participants will care more about one kind of process than another.

Cities are where people make connections and build relations of many kinds: in that sense, they are sites of serendipitous engagement. We call this productive urban serendipity "propinquity": socially, economically, or politically useful proximity. Cities are also sites of density, though what counts as dense is often perceived in relation to what surrounds the center of a cluster (see box, "Urban and the City"). Levels of density seen at the suburban fringes of New York, United States, or Manila, Philippines, would be characterized as central and bustling in more diffuse cities such as Brisbane, Australia, or Oklahoma City, United States.

Together, these two terms (propinquity and density) can serve as an initial shorthand definition of "the city." Yet we would also say that these two terms on their own are also somewhat unsatisfying. They help distinguish cities from their surrounding rural contexts by capturing the idea that cities contain something like "lots at once." In other words, cities have people, buildings, transportation arteries, and activities in relatively close proximity. But propinquity and density do less to explain the urban process that produces those cities or the character of cities as an object of analysis. *Why* are cities sites of propinquity and density? How do they become so, and how are the conditions of propinquity and density sustained over time? What motivates individuals to live in cities, and why do they stay? These and similar questions form the core of urban analysis.

This book attempts to cultivate some habits of mind that can make effective analysis of cities easier. The text explores how the authors, as urban scholars and observers, attempt to deal with the challenge of thinking about cities in a rigorous way while embracing their complexity and plurality.

Places and place-making

We advocate for an analytical approach drawn from the discipline of geography that focuses on *places*. Traditionally, geographers emphasize that places are made up of the

Urban and the City

The word "urban" captures the idea of a built-up area of human settlement. It usually also connotes density, distinguishing built-up areas that are relatively small, like mining settlements or small towns, from areas that have a significant concentration of human-built infrastructure (roads, buildings, and utility systems like sewers and power lines) as well as people.

A "city" represents a built-up urban place that is sufficiently large to need a political system that enables regulation of people and their circulation within that urban space. A city almost always includes a legal entity, with specific laws that enable a particular system of government to administer it. The laws creating city governments, and the sorts of powers that these governments have, vary by region and country, and so the details depend on where the city is located. Even if a city government is clearly established over an urban area, this does not mean that the city boundaries encompass all the urbanized land in the area. There may be towns or other built-up spaces that do not function politically in the same way as the city; these areas are urban but may be governed, or managed, as part of separate cities or towns, or as parts of a larger political region (which may be rural as well as urban). For example, many urban areas have adjacent but politically distinct municipalities that include built-up urban areas. Conversely, some cities contain areas of land with very low-density development.

So while city and urban are not the same thing, they are related. Cities always have some, if not completely, urban characteristics – namely density of people and built environment, and diversity of activities – but not all urban places are legally cities. Sometimes – often, in fact – an urban area contains multiple cities and/or towns (which usually have different political powers than cities).

Geographers, sociologists, and other scholars have explored "the urban question" of how much density of people and built infrastructure is enough to denote meanings that people attach to geographies, human processes that shape the landscape, physical characteristics in the environment, and non-human processes, all at once, iteratively changing over time. This understanding of place tends to foster a notion of places as singular agglomerations: a product of the accumulated history of people making geographically located meaning.

Doreen Massey was a geographer whose scholarly writing on places, economies, and cities has been very influential on the thinking of your authors here. Massey writes that places are bundles of space–time trajectories (2005). The word *bundle* has metaphorical connotations: while it has a literal descriptive meaning as a cluster of things bound together, its usage is often associated with a group of long things (sticks, wheat, etc.) bound up as a set with a ribbon or rope by a person. The use of the word "bundle" here seems to bridge its literal and metaphorical meanings.

A place bundle isn't like wheat in that you can't pick it up and move it elsewhere; but neither is it a natural, preordained set. Bundling describes the process by which people select or choose what counts in their own minds as a part of a place, and what does not. Bundling is an activity we're all engaged in, all of the time: making sense of the world by sorting it into this place and that place, according to our own understandings and logics.

But what in the world is a space–time trajectory? Massey tries to capture the idea that both physical and social objects are part of places. What does this mean? Think of famous plazas like Red Square in Moscow, or the National Mall in Washington, DC. These places include objects like sets of large buildings (the Kremlin, the Smithsonian Institution), as well as open pedestrian areas that most of the time contain little or nothing at all. But Massey also emphasizes that place bundles include social patterns and human beliefs or intentions. So the rhythmic return of US presidents to the Mall every four years to be inaugurated is a part of the bundle that makes up that place; so is the repeated pulse of bodies to celebrate the nation with fireworks each 4th of July, and its irregular occupation "urbanness," and what urban life is for and about (e.g., Wirth 1938; Castells 1977 [1972]; Brenner 2000). Early sociological ideas in Europe about urban life focused on the idea that people could be unknown to one another in a city, and thus were not bound by kinship ties and social norms as they had been in more rural contexts (Tönnies 1955 [1887]; Durkheim 1997 [1893]). These scholars also debated whether urban life was alienating people from their mutual obligations to one another. Manuel Castells, writing in the 1970s (Castells 1977 [1972]), emphasized the importance of social organization and government provision of infrastructure in urban areas, which supported communities through services such as schools, garbage collection, and roads. He also wrote extensively about collective organizing as a positive element of urban life (Castells 1983).

A crucial argument throughout much urban scholarship over the last century is that urban places generate new social ideas and change because of the bringing together of different people and resources in one relatively dense place or area. Geographer Ed Soja (2000) wrote about this function by pointing to the visual art on the rock walls of the ancient ruins of Çatal Hüyük, which existed in Turkey about 10,000 years ago. His point was that when people live together, they create more than just what they need to live. Other scholars have likewise highlighted the importance of social organization and expression as part of urban and city life. These include Ed Soja with his attention to arts in ancient cities, Louis Wirth (1938) and his description of the "theater" of urban life in Chicago in the 1920s and 1930s, and Henri Lefebvre (1996), who wrote about the *oeuvre* (connoting artistic and non-waged work) of urban residents.

by protesters of various kinds. The paths traced by those people through space and time (i.e., space–time trajectories) are transient, but their cumulative repetition in physical spaces makes the trajectories durable in people's imaginations of the place.

Place framing

Space–time trajectories can be social or physical or emotional or experiential, but the bundling process is always a human act, not a natural event. In previous writing, we have emphasized that trying to convince others that one's place bundle is the right place bundle is always in some way a political act (Martin 2003b; Pierce and Martin 2015). When a North American or Australian university makes a statement that it is situated on unceded Indigenous lands, it is attempting to persuade others about its relevant geographical context. Situating the university within a tribal geography as opposed to that of a regional state or a city has political implications. Yet so does the more conventional bundling into the geography articulated by the land's contemporary rulers, even if agreement with dominant norms raises fewer eyebrows.

Any effort to persuade people to share a geographical imagination is a kind of politics. We borrow the vocabulary of framing from social movement theory in calling these political efforts to persuade others of the correctness of one's own place vision as "place framing."

People are constantly jostling with each other about what constitutes a place, asserting that their own version is the correct one. Developers, city officials, and groups of residents, for example, advocate for certain forms of urban development as important and vital for urban life, while other buildings or businesses or even people are abandoned or left behind. We call these competing views of experienced geographies "place frames" because of how they focus attention on some geographies more than others, just as a window frame shapes a mental image about a landscape beyond the walls of a home.

As noted at the outset, urban places share the core characteristics of propinquity and density. They are all cities, after all. But different place frames emphasize different elements of the city, even as they might also include elements that are shared. Our argument is that one can't assume that the same characteristics of urban places matter most across different

History of Place Theory

Place as a concept often implies rootedness or location. A place *is* because it has a name, a meaning (multiple meanings) for some aspect of human life. The idea of a fundamental human connection to place was articulated in the 1970s by the geographer Yi-Fu Tuan, who wrote both that place is "a center of meaning constructed by experience" and also that places "are points in a spatial system" (Tuan 1975: 152). In this characterization, Tuan highlighted the difference between a focus on meaning and a focus on generalizable knowledge. He pointed to cities as particularly significant places in part because they are centers of population, where many individual experiences accumulate into sites of multiple meanings, even if these meanings are not all publicly evident in the concatenation of people, building, roadways, and activities that constitute any given city. More recently, Tim Cresswell (2014) has written about the importance of places in this tradition of located, meaning-laden geographical experience (see also Relph 1976; Adams, Hoelscher, and Till 2001). Extending Tuan's idea of place as experience, some geographers, environmental psychologists, and others emphasize human connections to place as part of an individual's place identity (Lewicka 2008; Kaplan and Recoquillon 2014; Devine-Wright 2015; Main and Sandoval 2015). This approach to place focuses particularly on individual cognition and emotion, as well as collective experiences, and how these shape place conception and understanding, as well as a sense of self or community.

Other theorists of place such as John Agnew (1987, 1989) and Doreen Massey (1984, 1991, 2005) understand it as multifaceted and relational. They draw on Tuan's and others' focus on individual experiences and cognition, while also emphasizing societal systems and processes. Agnew (1987) argued that place is comprised of three intersecting elements: locale, the immediate setting for everyday life and interactions; location, which is the broader economic, social, contexts because "what matters" is a matter of politics and perspective.

There is a kind of double movement going on here. On the one hand, people argue for a particular place by engaging in framing and persuasion; on the other hand, they sometimes claim that the place they are framing is natural, deeply rooted, and unchanging. This doesn't mean that people are disingenuous; it just means that people disagree about what is essential and timeless and what is transitory or illusory. (It is actually a quite common tactic for people to try to persuade others of their beliefs by characterizing those beliefs as natural or uncontroversial.)

Cities contain many different things and processes at once: for example, built environments, types of living places, and modes of living. Place framing is one of those kinds of processes, but climate change and tectonic shifts are examples of other (more physical) processes that also contribute to the city as a place. Because of the inherent complexity of cities, it is naturally a difficult task to decide where to enter and how to scope urban analyses.

We emphasize that urban places are made up of many different kinds of parts, and that these parts function both independently and also as part of a whole. Think, for example, of a stream that is part of a large urban watershed. On one hand, that stream is a resource that people care about for aesthetic, economic, and health reasons: it has multiple dimensions of value for different groups of people. On the other hand, it is also simply the result of processes of water cycling interacting with the landscape, insensate and indifferent to the social or political machinations of the humans that wander around nearby. The stream is both of the urban assemblage, a part of it; and also apart from it.

Thinking this way sounds easier than it is. If you practice seeing cities this way – as sites where many plural place frames overlap, where different actors compete to affect what "counts" and what does not – we think you will be able to see and understand what happens in cities more accurately and perhaps with more awareness of the limits of one's ability to see the city as a whole.

> and political situation of the locale in relation to other places; and sense of place, which captures the emotive, affective relation of site to individual. This three-part definition brings together the specificities of any given location, regardless of its scale, the interconnections of places, and the human meaning brought to bear on given geographies. Massey (2005) argued that places are in fact "bundles" of assembled ideas, material items, and experiences, all drawn from multiple scales and people. It is important to note, as geographers have (e.g., Tuan 1975; Massey 1991), that places are not necessarily small and intimate; one's kitchen is a place, cities are places, and so too are countries. Taken together, place scholarship tends to emphasize the meaning-ladenness of place for human experience, and the mutually constitutive relationship between place and people (both individual and collective).

Using a place-framing approach to think about cities

Thinking about cities as a series of overlapping, contested, and divergent place frames emphasizes that some elements of the urban context are foregrounded and that others are latent or intentionally obscured. A place frame is a lens on a city; the place-framing approach is a way to focus on the ways that different actors promote different elements of a city to their own ends.

Saying cities are places, as we do, is saying that the way to study cities is to examine their many constituent place frames. These different frames emphasize different elements: built and natural environments, modes of life, patterns of cultural celebration, and economic logics. Many scholars of cities focus less on cities-as-places and more on a particular way of explaining cities-as-processes; in our view, such approaches tend to offer singular place frames that prioritize specific elements of cities and treat those elements as though they function as a whole explanation.

Seeing the city first and foremost as a machine for economic growth (Logan and Molotch 1987) is a common analytic metaphor. That lens on the city helps to explain a host of processes oriented to production and exchange. It is natural to try to extend one's understanding by deepening or widening an analytic approach that has been helpful; this is a human tendency. We are sympathetic. At the far end of this tendency, however, you end up insisting that the meaning-laden geographies of others don't *really* matter – that their meaning making is explained by factors beyond their own understanding or focus. For example, Marxian urbanists often say that other modes of difference (like race or gender) are relevant to analysis, but that they are important *within* the Marxian frame of capitalist expansionism and class conflict. Conversely, while urban economists think that Marxian urbanists are wrong about what ails cities, they agree that economic processes are the decisive ones: cultural process or social negotiation are derivatives of underlying economic factors.

The economic is not the only possible analytical lens: feminist urbanists tend to focus on how cities manifest a series of male-oriented and male-dominated perspectives, such as how paid work is organized, how childcare is valued and spatially accommodated, and how houses are designed and constructed. Other systems that contribute to these elements (such as transportation systems or public health infrastructure) are important, but from a feminist analytical perspective they are first shaped by gender and then secondarily also impact social and economic systems. Committing to a single key analytical approach creates incentives to reinforce the centrality of that approach incrementally as analysis continues, *even when that runs counter to what residents themselves think matters about the cities they live in.*

Urban economists are not wrong that urban economies exist: indeed, these scholars describe and highlight important urban processes. But their descriptions are also inevitably incomplete. Too often, they see the part as the whole and, we argue, foreclose understandings that help us see cities as

Conceptual Framing and Place Frames

The concept of "framing" characterizes the interpretive schema (Goffman 1974; Giddens 1976) that "function to organize experience and guide action" (Snow et al. 1986: 464). Framing was originally articulated as a way of describing that individuals mentally select and order their ideas around a topic or schema, but it has also been understood in a more collective sense to describe social meaning and action (Schön 1980; Snow et al. 1986).

Scholars of activism, also known as social movement theorists, found this term particularly useful for understanding how people come to share and articulate agendas for collective action and activism. Sociologists David Snow and Robert Benford coined the term "collective action frames" to describe how groups of people come to share perceptions, grievances, and hopes for the future in order to advocate for social change (Snow et al. 1986; Snow and Benford 1992; Benford 1993). As Snow et al. (1986: 466) argue, political actors such as social movements "not only act upon the world, [. . .] they also frame the world in which they are acting." The concept has been extended to describe articulations of collective identity (Gamson 1992; Hunt, Benford, and Snow 1994; Polletta and Jasper 2001; Fominaya 2010; Adler 2012), as well as to understand dominant ways of knowing around social issues such as energy policy (Eaton, Gasteyer, and Busch 2014; Hazboun et al. 2019; Wolsink 2020).

Geographers Deborah Martin (2003b) and Hilda Kurtz (2003) each adopted the concept of frames to understand the ways that some social movements connected their grievances and advocacy about social change in reference to specific locations and geographies. Martin argued that activism oriented to improving the economy and landscape of a particular neighborhood demonstrated "place framing"; while Kurtz described an activist strategy of "scale framing" in environmental justice conflicts in Louisiana, United States. Both of these scholars argued that specific geographical contexts of activism inform collective organizing.

places not just more complex in the present but holding more potential place-ness (experiences, built environments, etc.) in the future. What we are saying is that to understand cities is to understand that they are always more than one thing at once. A city is many different machines for many different functions. If one understands a city as only one machine, producing one kind of outcome, then one has fundamentally misunderstood that city.

Our goal is to explore different understandings of cities in order to understand dominant meanings but, also, the ways that other meanings intrude and challenge, and what difference the framings make for urban life and possibility. Historically, seeing cities as sites of disorder and moral vice has been a common framing technique: here, the key characteristic of the city is that it is alienating, socially corrosive, and thus corrupting of the people who live there. Related to this is a moralizing frame of the city as a site of immigration and social difference: here, the city is a kind of frontier beach upon which migrants wash up with uncivil habits that break the social order. Another (perhaps contradictory) frame is of the city – especially the "global" city – as a kind of cosmopolitan playground for the privileged and wealthy. Each of these frames has adherents, both academic and non-academic; these frames tend to powerfully shape those adherents' understandings of what they observe in those cities.

Place politics

The process of urban place framing itself is negotiative, but framing as a process does not have a specific political or ethical valence. This means that while a given observer may see particular frames as having desirable (or problematic) ethics, framing itself is a process in which people of all political stripes participate. Place framing is intrinsic to social negotiation: it is a thing that all social actors do as they try to make the world they live in reflect their values and preferences.

> Hewing closely to the articulations of collective action frames, place frames include motivational elements that define and characterize a place-based community, as well as identifying grievances, or problems, that framers have identified and wish to solve with particular action or policy (Martin 2003b). Place frames thus offer a way of understanding how people think about and articulate shared norms or agendas for places, which helps to produce places both discursively and materially. Indeed, Martin (2003b: 731) argued that place framing works to "constitute places and polities at a number of spatial scales."
>
> Other geographers use the place-framing concept to explore conflicts over land use, socio-technical change, community organizational agendas, and urban policy generally (Larsen 2004; Elwood 2006; Pierce, Martin, and Murphy 2011; Uitermark, Nicholls, and Loopmans 2012; Murphy 2015). Pierce, Martin, and Murphy (2011) in particular propose application of the concept beyond collective action, arguing that place frames capture the ongoing active contestations of all places.

Emphasizing that framing is not in itself "good" or "bad" as a process makes it analytically easier to attend to multiple simultaneous frames by subtly changing the incentives of urban analysis. If an analyst takes (for example) a more conventional economic perspective on urban growth, there is an incentive to seek to explain as much of the urban landscape as possible in terms of the theory that has been adopted: more explanatory leverage means that your work has more power and predicts more. But if an analyst tries to understand how much of the urban landscape is explained by the application of Frame A as opposed to Frame B, the analyst is already oriented toward the limits of that frame: how much is A and how much is B. Starting analysis by focusing on the negotiations between advocates of multiple frames tends to encourage a more open-ended analytic approach.

Relatedly, an urban place-framing approach makes it easier to see and hear the voices and analytical perspectives of people who have historically been less empowered in the urban landscape. One widely agreed characteristic of the lives of poorer people, for example, is that a larger proportion of their collective "capital" is social rather than economic. Yet classic gentrification theory still often centers the ways in which marginalized urban communities are economically marginalized by rounds of displacement from neighborhoods (Smith 2005 [1996]; Lees, Slater, and Wyly 2013). An urban place-framing approach to analysis doesn't argue that economic factors don't shape the housing landscape – they do. Rather, it emphasizes that the reasons that poorer people mourn displacement are often not, in their own terms, centrally about wealth accumulation. Gentrification is as much a conflict of values about urban places as it is a conflict about the accumulation of wealth (Davidson 2009).

The focus of a place-framing approach to urban analysis

There are three important attributes of urban analysis centered on place framing that we seek to emphasize in this book. These are not the only characteristics of an urban place-framing approach, but we see these three as clearly in contrast with tendencies in the wider urban literature.

First: a place-framing-oriented approach to thinking about cities emphasizes a balanced approach to the effects of human agency and social, political, and economic structures. Social scientists use the word "agency" to describe the capacity of people to choose their own paths forward. As we've described above, scholars and public policy actors work to simplify urban analysis by reducing the number of factors to which they pay attention. It is often the case (though not always!) that in seeking to reduce those factors, they make it easier to predict the change of one factor in terms of one or a few others. For example, it is common in urban economics to celebrate an approach to analysis

Structure and Agency

In classical economics, the "rational actor" is imagined as a reasonable, completely self-actualized individual who makes an unending set of free choices to their own maximal benefit. In contrast, scholars in political economic traditions have tended to emphasize our capture within social and economic structures that make it difficult to even imagine alternative pathways through life (Harvey 1989b). This distinction between agentic and structural explanations was, for a time, a point of significant conflict between social scientists.

Pierre Bourdieu (1977) and Anthony Giddens (1984) were key theorists who attempted to reconcile structure and agency, saying that humans are both structured and agentic. Giddens used the word *structuration* to describe the mutually constitutive processes through which structures shape people and people shape structures. Since the 1990s, most urban theorists implicitly accept that both agency and structure shape societies, while varying in the degree of emphasis on one or the other.

In examining agency, feminist scholars such as Iris Marion Young (1990) and J. K. Gibson-Graham (1996) examine the ways in which people create their own subcommunities in cities, and can develop bonds of support and exchange that transcend or work around the dominant forces of money-based exchange. Sociologists Marcus Anthony Hunter and Zandria Robinson (2018) detail the many ways that Black Americans have shaped the culture and community infrastructures of US cities, despite institutional racism that mostly ignores these identities and support networks, particularly in scholarly accounts of urban process and culture.

Theorists who focus more substantively on structures highlight the role of the economic system in shaping urban economies based on property value and exchange (Harvey 1989b; Blomley 2003). In such accounts, the economic mandates of capital investment drive urban processes from that "explains" 30 percent of a variable's change (like, say, how vertically a city is built) based on (say) only four other variables. Urban political economists, similarly, may see agency as a real phenomenon but often see it as overwhelmed by the structural dynamics of modern capitalism, on which they focus.

We agree with these actors that structures exist and that people's individual choices are often influenced by wider processes. But we also see that people are constantly making choices about the places where they live. Those choices (often small, everyday ones) iteratively produce and reproduce cities over time. In paying more attention to agency, we push back against the strong tendency to center economic explanations for personal choices at the expense of others, even though economic factors matter. In this sense, we think that focusing special attention on agency is a corrective toward analytical balance and integration of structure and agency, rather than a claim that societal structures aren't present or don't matter.

Second, thinking about cities as places emphasizes that all understandings of those places are *partial or incomplete*. When people articulate place frames, they are always, inherently, including trajectories they want to emphasize and excluding others which undermine their narratives. If an individual frames Baltimore, MD, United States, in terms of the centrality of its predominantly Black population, it might downplay the city's long history as a home to one of the largest Jewish populations in the United States outside of New York City. Alternatively, if a community frames Baltimore in terms of the long history of its working waterfront, they may center the idea of Baltimore as a distinct urban-economic region, downplaying its economic and political interconnectedness with Washington, DC.

Focusing on place bundling and place framing doesn't mean that one's goal is to expose *all* of a city's place frames and synthesize them into some sort of mega-frame. Instead, we take note of how analytical and/or political goals shape bundling and framing. One attends to the trajectories that are needed to achieve an understanding toward an end.

> what is built to where it is built to why municipal authorities help developers to obtain property for private investment. So while more structural accounts acknowledge a role for individual human agents, such as city officials or developers, their actions are understood to be shaped by the demands of a broader system, usually capitalism, rather than by individual choice or understandings.

This is true of scholarly analysis, political analysis, and the analysis of supposedly apolitical residents who merely want to protect the mature trees on their streets. Everyone frames, and frames are always partial.

Third, a place-framing-oriented approach to thinking about cities emphasizes that place frames aren't just plural, or multiple; they are *radically plural*. Because everyone engages in bundling and framing, attending to place means seeing the world as full of overlapping places. Lots of people will sort of agree with each other about key elements of what constitutes Aberdeen or Birmingham in the United Kingdom; that agreement represents effective place framing over time. Yet close examination will reveal that their frames actually diverge extensively on (say) whether the nearby oceans are a part of Aberdeen, or (say) whether the expansion of high-speed rail is slowly embedding Birmingham within the placeshed of London. These disagreements signal overlapping, not thoroughly congruent, frames.

Thinking of cities as radically plural – as sites of multiplicity, of place bundles sprouting endlessly in every direction as life reorients residents to recognize their cities anew – orients one toward thinking about the problems that specific place bundles and frames solve in people's lives, rather than toward some one-true-theory of the logic of the city. When people reorient their urban lives around childcare, or petcare, or anti-war politics, what the city's ways and stations enable or obstruct for them suddenly changes. The city is a (metaphorical) machine for living, but what constitutes *living* is always in flux. So too, then, is the city.

Infinitely branching (subjective) universes: exploring urban place imaginaries

Cities are made up of many different kinds of elements: for example, buildings, people, communities, identities, and future goals. In urban contexts, these elements are densely, cacophonously packed in. Each element has its own individual trajectory (a car tracing a curve through an on-ramp onto a highway, or a high-school student whose life is headed toward achievements and an office in a tower high above crowds on the street) but, when combined, these parts constitute an object that is more than the sum of its parts. In thinking about the "big" elements of a city – such as its economy, cultures, built environment, natural environment, history, or housing – a separate story could be told about each object or agenda. But so too can stories be told about how these elements are combined.

Imagine a person walking along a sidewalk on a city street, perhaps a downtown street. Depending on where in the world this downtown street is located, the specific sights – the construction of the sidewalk, the width of the travel lanes, the affordances for pedestrians, etc. – will vary. Nonetheless, the street offers myriad sights, sounds, and smells. There are other people walking along the street singly, in pairs, and in the occasional group, and a few others stand in doorways. They are wearing a range of business and casual attire – suits, formal dresses or skirts, jeans, and in some cases worn or ripped clothing – and some are carrying briefcases or looking at phones. Some people tear past the walker, rushing ahead to the next intersection. Others stroll more casually. Snippets of conversation are audible, laughing as well as more strident tones. In one doorway, a person is sitting on a stoop, hair unkempt, wearing stained and worn clothing, a large plastic bag beside them. A waft of body odor comes from the spot. Along the street are buildings of concrete, glass, and brick. Signs large and small in a mixture of languages indicate banks, restaurants, coffee shops, and law and accountancy offices. Some people are visible in large street-level windows at a bank: the windows draw the eye inward, with the tellers' counter and tall writing desks visible from outside. In the street, a taxi, bicyclist, and a bus each pass in the lane by the sidewalk, negotiating the space tacitly with each other and other cars in the street. Engines rev as a light turns green; a horn blast echoes from a block away.

There is more to observe here than any observer can process. Any participant in such a scene – even a viewer of a photograph – will inevitably focus more on some sights (or if in person, sounds and smells) than others. Each element of this bit of a downtown place has itself a complex story of functions and why they are there, in that spot at that time. A bank, for instance, provides a location for individual customers to access cash, deposit checks, or engage in other financial transactions. The bank provides similar services to a variety of businesses located in this city. But it also has other functions related to its role in an economic system: lending money to businesses, for example, including real-estate developers who might transform some of the nearby landscape with those funds. People will go into a downtown bank for a variety of reasons; to work there, to meet a loan officer, to open an account or deposit/withdraw money from an existing account. The people who work at the bank perform a variety of functions: security guard, cleaner, teller, loan officer, bank manager. Each of them has a presentation, a work "uniform" and a set of actions they perform that help to distinguish their roles in that place. In order to get a job at the bank, each person would have had a different social, educational, and training path to get there.

An observer in the bank at that time might notice some workers more than others, making different assumptions about the job or the person based on superficial appraisal of clothing or visible bodily identity that may not at all, or only in small ways, match that person's own experience of the world. These cursory judgments or assumptions might be largely influenced by one's own reason for being in that time and space, and needs or desires for interactions as simple as "hello" and as complex as accessing funds in an account or paperwork in a safe deposit box.

Turning away from the bank, consider the bus that passed by. The people on the bus – maybe it was about half full – are on it for a variety of reasons. Some of them are headed to work: perhaps at other banks, or a restaurant, or in an office tower working as receptionist for an accounting office also located downtown. Others on the bus aren't going anywhere downtown, but had to take this bus from a stop near their apartment building in one part of the city to the downtown bus station in order to get on a different bus that will get them to work in a retail location. There is no other way to get from one part of the city to the other without going through downtown for this person who takes a bus to work rather than a car or a bike. In some cases, that is a choice to use public transportation, but for other people on the bus it's a necessity to take the bus because they can't afford to buy and maintain a car – or because driving in a personal car in this area is restricted, too expensive, or simply too slow due to traffic. Other alternatives, such as biking or walking, might not work, due to the distance or the weather.

Thinking about the people and places you see downtown in a city, there are many stories you could follow and understand about why people are there, or why businesses are there. The same is true for residential areas of a city or business districts outside of the downtown. But in a residential area or a business district, you would likely see less variety of functions, building types, or people as you would see downtown. That variegation depends on the size of the city and where it is located in the world.

Thinking about the city overall, there are many ways to understand what comprises it, and why the businesses and activities and people in that city are there. Urbanists understand cities as a mixture of people, functions, and activities in relatively limited space, and try to generalize about the functions, activities, and experiences of cities (their serendipitous propinquity). The challenge for seeing cities in all their complexities lies in developing a framework that is attentive to multiple, intersecting, and sometimes conflicting dynamics. The framework we advocate here starts with urban places, that is, with cities as places with many simultaneous stories and elements. It allows us to notice how different slices of that complexity imply distinct politics and distinct visions. Thinking about different slices suggests different ideas of what a city should be and how different people fit into – or are excluded by – those visions. The overarching framework that helps us to do this work of recognizing complexity, while not being overwhelmed by it, is to think of cities as places.

It is hard to think about an urban place as many different kinds of object all at once. It is possible to acknowledge superficially that these different elements all exist at once, but much harder to treat them all as simultaneously present and true. A place-framing approach to analysis helps to do this work.

A place-framing approach acknowledges and deploys many different and simultaneous lenses by incorporating them into an understanding of cities as patchworks of many different local and global entities and relationships. A relational understanding of place-making starts with an idea of the urban as a site of life: it attends to human experiences (singular and plural) and social meanings, while also understanding places as produced through a series of related yet distinct political, economic, cultural, and social processes and practices. Considering how frames capture different aspects of cities helps us to remember the small detail of a person sleeping in a doorway. Thinking about urban places as relational means paying attention to what is visible and explainable through a particular place frame; to then think about what we are missing; then to shift to a different place frame, and in doing so to think about the diversity within and complexity of the urban place under examination.

The book that follows: examples of differential place framing in cities around the world

Each of the six case study chapters of this book will illustrate how different urban frames highlight different ways of thinking about urban places. Each chapter looks at a different city, with our examples drawn from a broad range of sizes of cities in very different contexts. The cities have different kinds of political economies, cultural characteristics, and distinct urban histories on wildly varying timescales. We have selected these cases to help you to develop the habits of mind to think about urban places as relational, plural, and made up of overlapping place frames. Across the case study chapters, we've chosen some frames to emphasize that they are supported by constituencies with different levels of power. Some of these frames might be characterized as dominant, while other frames are less common or even marginal, but are still important in explaining the ways that urban processes proceed for different communities in different cities. The frames we've highlighted certainly do not cover the whole terrain of the politically possible in urban place framing, but they do offer some breadth of perspective and pointers for how to explore alternative understandings of cities. Each of the case study chapters also includes critical vocabulary in set-apart boxes to help orient the reader to key concepts in urban studies and research.

You have nearly completed chapter 1, the Introduction: this chapter introduces our basic analytical approach and introduces some vocabulary that will re-emerge periodically along the way.

Chapter 2 examines the City of London, a small historic enclave within Greater London in the United Kingdom. The chapter explores the interaction of two place frames: one as an ancient shelter *by and for capital* and the other as a *vanguard for a modernist urbanism*. These two frames aren't in conflict with each other, exactly, in the London of the present day; but they have distinct advocates, and they imagine the City toward different future trajectories. In telling the story of the City of London, we also explore the history of urban capital and the central role of political economy in much of contemporary urban theory.

Chapter 3 examines Tehran, the capital of Iran and its political and economic center. We explore Tehran's status as a city in the Global South, and the ways that its leaders over the last century have focused on a developmental frame emphasizing infrastructure for economic activity. As a result, one dominant place frame in Tehran is *economic intensity and growth*, and we examine the growth coalition of actors that benefits from this orientation. At the same time, the city also represents a *cosmopolitan site of diverse modes of life*, where residents express a variety of lifestyle choices within a seemingly strict Islamic framework, and use the city's openness to diversity to push for greater input in the infrastructures of daily life in residential neighborhoods.

Chapter 4 examines Worcester, a small city in Massachusetts, United States. As a smaller city in a region dominated by a larger city (Boston), Worcester represents the challenges of local decision making for charting urban identity. The city has struggled to exert local autonomy in two distinct realms: downtown development, and environmental crisis response. In the first framing, Worcester's local officials have sought to continuously remake *downtown as a regional economic center* as a response to postindustrial economic restructuring. The second frame, *the city as a place for nature*, highlights the ways that residents rejected "outsider" expertise from federal and state officials when an invasive pest was discovered, affecting a portion of the city's trees. In both these narratives, the city is buffeted by outside forces, whether these are economic, environmental, or political, and these two frames illustrate the uneasy coexistence of local identity and regional political economy.

Chapter 5 examines Portland, a mid-sized city in Oregon, United States. Portland has been seen for several decades as a left-leaning *ecomodernist urban paradise*. But the city-region has a longer history of state-sanctioned *racial and colonial erasure*. Because of the left-leaning politics of the city, the recent eruption of protests against racial discrimination and police abuse have to some degree destabilized the hegemony of the ecomodernist frame, opening up ambiguity and discomfort for residents about the core identity of the city and its region.

Chapter 6 examines Chongqing, a key industrial city in the Chinese interior, with much of its built environment situated on hills where two rivers meet. Its framings reflect the influence of social media on city imaginaries, as Chinese and international tourists alike have celebrated the city as *a cyberpunk "strange" built environment*, with buildings built into hillsides. At the same time, its location in southwestern China has made it politically important as a site for experimentation of more *egalitarian urban development*, balancing growth with greater protection of rural areas and livelihoods.

Chapter 7 examines Jerusalem, an ancient city at the eastern edge of modern-day Israel. We illustrate how urban visitors to Jerusalem experience snapshots of an urban place, selections that to a large degree are curated based on the tourists' (often religious) demands. Jerusalem's Old City in this framing is *an inclusive tourist destination* of significance for three religions. Jerusalem means different things to different groups of residents and tourists alike, and thus offers a contested, complex multiplicity of meanings that are commodified as religious iconography through tourism. An alternative place frame is to understand Jerusalem as a state project that legitimates and cements Israel's identity with *Jerusalem as the capital of a specifically Jewish state* in conflict with its Palestinian residents, some of whom are also Israeli citizens.

Chapter 8 offers a conclusion about thinking about cities. We bring together threads from chapters along the way to make an argument for why urban place-making offers a productive alternative to other ways of organizing your thinking about cities.

As you read the individual case study chapters, consider how they offer distinct, place-oriented understandings of the cities they describe while also pointing to broad themes that recur across the chapters. A place-framing approach to cities constantly shifts between describing pervasive, dominant narratives and uncovering less explicit yet important dimensions of places that center other perspectives and experiences. The structure of this book and its chapter ordering does some of this shifting as well, starting with a well-worn case, London, yet focusing quite narrowly on a less exposed and often less understood element of it (just one square mile).

–2–
City of London: A Machine for Living/The Seat of Wealth

The City of London sits at the heart of the city of London. That sentence is not in error: the City (with a capital "C") is a roughly one-square-mile-sized area at the center of the wider London city-region. The City is the wealthiest of London's major business districts, home to a number of major firms specializing in international banking and corporate finance. Its prize jewel may be its foreign exchange market, which is the largest market of its kind in the world: nearly 40 percent of all global currency trades are executed there.

Despite its location at the center of metropolitan London, the City is a municipal enclave: that is, it is not administered by the wider city's government. The wider territory that includes most of the urbanized area is governed by the Greater London Authority. Thus the City is surrounded by (but not included within) the municipal government most people would recognize as "London."

The City is led not by the Mayor of London, but instead by the Lord Mayor of the City of London. Of the two, the Mayor of London has the far more important job in terms of contemporary influence and budget. However, the Lord Mayor's powers are rooted in the ancient history of England rather than contemporary legislation in the United Kingdom (as is the case for the non-Lord Mayor). In some senses, despite having a much smaller budget and much less influence over the day-to-day activities of human beings, the Lord Mayor's autonomy is also more absolute. Historically speaking, the office predates the British monarchy.

The City's governmental authority is invested in the City of London Corporation, a pre-medieval company or commune. That corporation's historic rights and privileges were affirmed by William the Conqueror in 1066 as long predating his ascension to the throne; this affirmation was a negotiated precondition for the City's submission to his rule.[1] The preeminence of City's "ancient liberties and free customs" were re-acknowledged and reaffirmed by King John when he signed the Magna Carta 150 years later, in 1215. (As of this writing, this affirmation of the prerogatives of the City is one of only a select few clauses from that famous document that remain in force as contemporary UK law.) Thus, while the Corporation of the City of London is not by any means an equal to the British Crown, its prerogatives are in a certain way prior to the Crown's. The reigning king or queen of the United Kingdom is recognized as the city's sovereign; yet when the Queen wants to enter the City, she must obtain the express, formal permission of the Lord Mayor before crossing into its streets.

The City sits upon the ruins of what was originally a Roman settlement that had a strategically useful location, but one of middling economic relevance in the context of the wider Roman Empire. The territory of the City (which is sometimes referred to by locals as "the square mile") is roughly bounded by the path of a Roman wall that was built in the second and third centuries CE. The settlement is thus centuries older than the country of England. Abandoned by the Romans and left for a time to decay, London became a thriving center of Saxon trade and wealth building perhaps 400 years after its founding. The idea of the City of London as a place for facilitating international trade (and as a privileged lockbox for the wealth that its traders might accumulate) was thus formulated in antiquity. This ancient history reverberates through this urban place into the present.

In the text below, we trace two place frames that different actors use to argue for the logic and function of the City. The first, and older, frame is the City as a place *by and for capital*. This frame builds on the City's ancient narratives of autonomy and trade with Europe and the wider world. Its proponents see the City of London as a kind of safe haven for buccaneering capitalists whose corporate wealth in turn entitles them to citizenship and even formal voting rights in the Corporation of the City of London, whether or not they reside within its bounds. The second frame – distinct from, though as we shall see not always in conflict with, the first one – articulates the City through the eyes of its very small complement of full-time residents, the overwhelming majority of whom have (for the past 50 years) lived in a single high-rise mixed-use development called the Barbican. Built by the Corporation in the years after World War II, the Barbican is a kind of high-modern idyll: a form of social housing built at the heart of modern capitalism for a cohort who needed no subsidy in order to live well. Its residents thus tend to frame their participation in the City as a kind of *vanguard for a modernist urbanism*.

In some ways, these two place frames both incorporate similar intuitions about British exceptionalism and London's global importance. The adoption of both frames certainly benefits a class of people with substantial wealth and privilege. But the frames imagine the trajectory of the City toward quite distinct ends. One (the vanguard) is modernist, rational, domestic, granular, and cosmopolitan; the other (for capital) is ancient, titanic, buccaneering, and globe-striding. These two frames, then, entail relatively distinct stories about what the City is for; its appropriately connected hinterlands; and its role in the project of a future for London and for Britain.

The City by and for capital

In order to make sense of the City as being by and for capital, first we have to understand what capital *is*. One way of thinking about capital is as "wealth, put to work" or "wealth in motion." Capital is wealth assigned to a task, which is to facilitate the organization of materials and labor so as to produce a return on its investment. Capital is thus a *mode* of wealth. To draw the distinction more sharply, a sum of wealth may be counted in dollars, pounds, or yen. However, if it is not used as a lever to organize the use of materials and the labor of humans toward the production of more capital, then that wealth is not strictly speaking capital.

This distinction between wealth and capital helps to understand the by and for capital place frame for the City of London. For a millennium, the City has been seen by its citizens as a dynamic, thriving hub for trade; but from the expansion of Britain as a global trading power and a colonial empire in the 1600s, the place identity of the City was increasingly drawn in contrast to two other distinct kinds of wealth within the kingdom.

The wealth of the City was often seen in contrast to the wealth-in-land of the landed aristocracy. The inheritance of feudal power and wealth through agrarian estates was seen by the traders of the City as increasingly dated and unimpressive. Additionally, however, the City was also to some degree seen in contrast with the wealth emerging from manufacturing and heavy industry during a period of rapid, fortune-building industrialization. While industrial manufacturing was *not* out of date – it, too, was understood as modern and advanced – the City was economically distinct, a place that was principally for the practices of trade and finance rather than production itself.[2] So it was simultaneously seen as distinct from wealth at rest *and* from wealth in motion as part of heavy industry.

From the 1600s through the 1800s, the City was instead increasingly seen as a place which motivated, coordinated, and facilitated the movement of capital through both national and global systems. For example, Lloyd's of London – the famous private insurance market – started informally as a place people could go looking for marine insurance in a coffee shop frequented by shipping agents in the City. During this period, the traders and financiers of the City of London developed an identity for themselves as desk- and

Capitalism and Urban Political Economy as a Comparison of Systems

Urban inequity is a significant theme in urban scholarship, dating back to Friedrich Engels' (2010 [1845]) observations on the abject poverty of textile workers in Manchester in the mid-nineteenth century. In the 1800s, what we now call economics was usually labeled political economy, emphasizing how political decisions are intertwined with structural and material aspects of production and consumption. In the contemporary era, this line of scholarship often focuses on critiquing the ways that various contemporary models of economic activity – broadly grouped under the heading of "late capitalism" – differentially contribute to inequality and human suffering.

Today, the discipline of political economy continues in this tradition. But from it the distinct discipline of economics diverged in the early-to-mid 1900s. Economists do not in general deny that political factors affect economic systems, but this discipline has moved toward quantitative modeling of economic systems *within* the contemporary "rules of the road" of liberal economies, rather than interrogating the rules and assumptions baked into a liberal, or neoliberal, consensus about how economies operate.

These two theoretically distinct threads of literature, then, characterize different underlying rationales for urban inequality. One thread, based on neoclassical economics, understands land value as a function of access – to decision making, to transit, and to production and markets. Any given location has a value in relation to what is located near it or accessible to it, based on what people, or companies, are willing to pay for the location (Alonso 1960). The second thread, more pervasive in contemporary urban studies, accepts the neoclassical logic of competition for land, but adds a Marxian approach by situating urban real-estate dynamics as actually driven by speculative investment (Harvey 1989a). Attention to speculation focuses analysis on the ways that land provides a site for investment for warehouse-based swashbucklers, taking financial risks in pursuit of accumulation that left some in ruin but others with commanding financial power. They moved (newly forming) financial markets with their choices. They offered credit to companies, saving them from collapse; they withdrew it from others, breaking those they disfavored.

During the overlap of the peak of mercantilism (roughly, the mid-to-late 1600s) and British colonial expansion toward the first British Empire (the late 1600s into the 1700s), British colonies were forbidden from trading with ports other than those in England. London became a – perhaps *the* – key hub in a global network of traders of spices, teas, furs, agricultural products, precious metals, textiles, and (most tragically) millions of enslaved Africans. From 1559 to 1786, it was illegal to offload international cargo anywhere in the London region except for 20 designated quays in the City, directly adjacent to London Bridge. As a result of the combination of a local legal monopoly on port activities and an English (later, British) monopoly on legal trade with the empire, the City was the overwhelming location for financial and commodity markets during these key years.

As other UK port cities (like Liverpool) grew in the 1800s, London's absolutely overwhelming position with regard to trade in the United Kingdom declined to a degree. However, the special role of the City of London as a globally crucial site for international finance expanded rapidly. Alongside the rapid growth of industrial capitalism in England, the City consolidated its position as a (perhaps the) preeminent center in Europe and the Americas for finance, especially trade-related finance.

Today, the by and for capital place frame is adopted by actors who perceive the standing of the City of London as the result of a set of values that guide the United Kingdom in the world. Those values include a developmentalist relationship with a global urban hinterland, an imagined centrality within that terrain, and a belief that London's power is rooted in its ability to set and leverage the terms of exchange for business transactions far beyond its walls. The by and for capital frame narrates the City as a kind of open-walled

> surplus capital (money not actively used in the sphere of production). People can and do "hold" land as an investment, even while not actively using it, and they might benefit from perceptions of low value during that time (so that their property taxes are low, for example).
>
> As locational factors change, investors can benefit. They can build and sell higher-value housing, which, in turn, makes the area more appealing to prospective residents who can pay a higher rent. These processes can lead to displacement of people who can't afford newer, higher rents; the term gentrification captures this dynamic and is a major focus of urban research, from underlying land dynamics to government-led, and social and ecological impacts (Hackworth and Smith 2001; Smith 2005 [1996]; Davidson 2009; Curran and Hamilton 2017; Anguelovski et al. 2019).
>
> *See also related box "Land Markets" in chapter 6.*

citadel, where dealmakers are protected in their transactions and from their interventions. This vision of the City of London is certainly British, but in a way that identifies true Britishness with the traits of the City as much as the other way around. The project of world making was once, in this view, a province of both the corporations of the city and the naval might of the nation. But where the nation has fallen away since World War II, the City's buccaneers have taken up the slack.

It is true that some of the leaders of the City's titanic, globe-striding corporations live near their work. Yet it is more typical to live much farther away, beyond the limits of the ancient City. Whether they commute from leafy outer suburbs or tony residential districts elsewhere in the core of the metropolitan region, the number of resident workers in the City is few. In the next section, we articulate the place frame not of the workers or the owners of the businesses of the City, but those who live in its sole major residential district: the Barbican Estate.

The City as a vanguard for modern(ist) urbanism: the Barbican Estate

During the half-century between 1800 and 1850, the residential population of the City of London stayed steady at roughly 130,000 within its one-square-mile area. Residents were stationed up and down the class hierarchy: trade merchants, traders, and bankers certainly lived near their work, but servants, staff, and local sellers also made their residences locally. (This historical pattern echoes that of cities around the world when the primary mode of transit was either walking or using draft animals to ride or pull a cart or carriage of some sort.) Then, over the following century, the population precipitously declined to nearly nothing, losing between 20 and 30 percent of its residents in each proceeding decade from 1861 to 1971. This ongoing residential loss was not a signal of a decline in the City's fortunes. Instead, it was an artifact of its status as one of the wealthiest business districts in the world. As rents increased and the spatial footprint of Greater London grew, the residential neighborhoods of the region steadily migrated outward from the City into other regions of the city.

The role of the City as a business district above all other uses was reinforced and consolidated by development strategies in the aftermath of World War II. The years-long bombing campaign of London by the German *Luftwaffe* left an urban landscape perforated by damaged and collapsed buildings. Like other major cities across Europe, London needed redevelopment plans for those sites. During this time, modernist theories of urban planning and design were ascendant. In the pursuit of rational separation of urban functions, cities across the world (but especially in North America and Europe) set about remaking the urban core so as to segregate different uses of space from each other.

Workers were increasingly expected to commute from residential estates in other parts of the urban region. Middle- and upper-middle-class workers increasingly sought housing in new suburban developments. An increasing percentage of

> **Urban Economic Processes**
>
> Urban economic theory focuses on the location as well as the circulation of people, money, and things. Urban economic scholars are interested in understanding why some cities have more (or less) growth than others, or why some areas within a given city have more (or less) growth. A major thread of urban economic scholarship explores how cities promote economic growth by acting as "entrepreneurs" (Harvey 1989b), seeking out businesses to locate within their territories that will provide jobs and a tax base (Molotch 1976). Scholars examine the wide range of actors engaged in promoting urban growth, including elected city officials, government and quasi-public development agencies, real-estate developers and speculators, and even non-profit foundations and organizations (Leitner 1990; Martin 2004; MacLeod 2011; McFarlane 2012).
>
> As economic development agents, city governments engage in a broad range of actions, from creating special tax-incentive districts (which typically offer developers or businesses large property tax breaks to locate in particular areas) to sponsoring "events," such as major sporting competitions or art festivals, that promote their cities as culturally significant sites (Ward 2006; Allen and Cochrane 2014; Lauermann 2018). City governments' role in spurring economic development means that they take economic risks, leveraging some developments by granting property tax breaks, for example, and participating in the complex financial arrangements which enable a global reach of capital investments in urban growth (Surborg, VanWynsberghe, and Wyly 2008; Davidson, Lukens, and Ward 2021). This financialization of urban policy has drawn considerable scrutiny from urban scholars, highlighting the broader risks to municipal government solvency, continued service delivery, and residential wellbeing (Rutland 2010; Weber 2010; Ponder and Omstedt 2019).
>
> The wide-ranging development activities of city officials mean that they act beyond their borders, as elected officials,

those in upper management had larger estates at the edge of the city. Slums concentrated and segregated lower-class housing. The increasingly default assumptions about the pattern of daily life – come to work in the center city, then turn outward to housing elsewhere at the workday's end – were formalized in the 1944 Greater London Plan, which called for massive construction of new suburban homes at the then-fringe of the London region.

However, there was also a desire on the part of some in the Corporation of the City of London for there to be a local housing option that would be desirable for high-level executives and their families. The motivations for this desire were varied. Some stakeholders were traditionalists who resisted the idea that the center city was no longer designed for them and their residential needs. Others had a specific personal desire for homes as close as possible to their corporate headquarters.

Additionally, at this point the number of full-time residents within the City's boundaries had dwindled into the mere thousands. Some local stakeholders were concerned that if virtually all residents left, it might induce Parliament to look more closely at curtailing or restructuring some of the City's fiercely protected historic rights and privileges. A small corner of the City near Cripplegate was thus set aside in the 1944 plan for something other than another large office building, though the site's exact purpose was at that point still up for debate.

Throughout the 1950s, the Barbican site (which had been systematically destroyed by bombing in 1940) lay fallow as advocates of different development plans fought to determine the outcome. A number of local landowners and builders were advocates for various private schemes to redevelop the site in a way that would be significantly more commercial in character, a shopping district with offices but only incidental residences. A separate power center within the Corporation, however, politically maneuvered for a large residential estate. This development would be densely settled, solidly built, and nearby to the corporate offices where imagined tenants would work. The advocates for residential development won out.

planners, and other urban policy actors engage with one another across regions or even globally, to network and share policy strategies (McCann and Ward 2011; Beal and Pinson 2014). The mobility of urban policies globally means that city officials face tremendous pressures, politically and economically, to participate in what seems like accepted wisdom for urban strategy, prioritizing actions learned from other cities, whether in the realm of sustainability, development, or smart city investments (Weber 2010; Clarke 2012).

Not all cities have the same opportunities for development. In the 1990s, sociologist Saskia Sassen argued that only some cities fully participate in the global economy. She identified "global cities" for what she called their command and control role in global economic forces, based in part on the presence of major stock markets where decisions about investment and economic trends are made (Sassen 2001 [1991]). In this theory, New York, London, and Tokyo are the most powerful urban economies. This work was influential in linking urban economies with intra-urban labor markets and polarization because of Sassen's attention not simply to decision-making elites but to the significance of service-oriented labor in cities and the wage gap between these two types of worker. Other research in this vein examines social divisions and inequality within cities (Allegra, Casaglia, and Rokem 2012; Modai-Snir and van Ham 2018).

See also related boxes "The Growth Coalition" and "Urban Agglomeration" in chapter 3; "Policy Mobilities" in chapter 4; "Residential Segregation and Redlining in America" in chapter 5; and "City Branding" and "Land Markets" in chapter 6.

In principle, this estate could have been privately developed through government condemnation and resale. However, actors within the Corporation wanted to retain more significant control over the path of development, and to do so they kept the new development in-house. The City thus structured its new housing development as many local councils at the time did: through a subsidiary governmental unit with some operational autonomy that nonetheless still was responsible to the Lord Mayor and the Common Council.

By the 1950s, modernist architecture and design principles were ascendant in the United Kingdom as in many urban places in the world. While some early drawings had more traditional facades, the firm that eventually produced the general integrated site plan and the specific building plans chose a late modern aesthetic: a palette of exposed concrete, long horizontal lines, and repeating geometric forms. Additionally, the functional approach of the site plan was in some ways prototypically late modern: the enormous multifunction site had its back to the street; finding your way into the main interior walkways and gardens seemed like a maze from the outside even when the logic seemed clearer once you were already inside.

From the inside, however, the Barbican seems to fulfill the promise of modern design in a way that few large-scale residential projects have. Concrete residential towers in the brutalist style have a wretched reputation, particularly those built in the 1960s and 1970s for lower-income communities. Tall towers with rough concrete exteriors and large courtyards disengaged from street life are today often associated with crime, poverty, and decay. Yet these spaces have elegance, proportion, and even a touch of spatial whimsy. The complicated paths throughout the complex lead pedestrians through unexpected juxtapositions and vistas onto flower-lined plazas. Residents play with the relatively uniform material palette and strong visual rhythms by decorating their balconies modestly, not interrupting these patterns.

Why is this complex of buildings a success where others of its era have failed and been demolished, sometimes after only a couple of decades? A key part of the story is that the Barbican was built from the beginning for middle- and upper-income residents. Despite superficially similar aesthetics, the buildings in the Barbican complex were built to a much higher standard and have also been sufficiently maintained as time has passed. For example, many tall "raw" concrete buildings are simply poured into rough wooden forms that,

Modernism and Urban Design

Urban planning emerged as a licensed profession and a clearly defined academic discipline in the United States and Europe in the first half of the 1900s (Hall 2014). This period aligned with a number of key global events: two world wars, during which the economies of Europe and to a lesser degree the United States were reoriented toward a project of "total war," and between them roughly a decade of economic depression. This period also aligned with the rise of modernism, an aesthetic and theoretical movement that emphasized human perception, rational powers, and a kind of technologism, which put faith in the rational application of science and technology to address human social needs and problems (Harvey 1990; Gold 2007). Modernism was a diverse movement, but it was underpinned by the idea that much of historical human practice was devoted to recreating patterns that were untested by science. It was an optimistic movement toward a better world made via careful and empirically supported rationality.

During the period from the late 1910s through the end of World War II, modernist principles became a key element of urban planning and architectural training. Yet because of intense war and economic depression, during many of these years very little investment was made in urban landscapes. At the end of World War II, large parts of many key European cities had been systematically destroyed by bombing. Thus the ideas of rational modern urban design, building for years without many practical opportunities for their application, were suddenly applied at a massive scale to the reconstruction of Europe and the suburbanization of the United States. This was a kind of experiment in urban design principles over several decades at a scale hitherto untried (Klasander 2005; Irwin 2019).

Modernist urbanism emphasized clear separation of uses (housing, commercial, transport, and industry); clean straight lines; large blocks of color and especially the clean, clinical feel of white surfaces; and scientific functionalism without

Residential balconies and gardens at the Barbican Estate[3]

unneeded, sentimental adornment. Technological advances in this period include curtain glass walls, the advancement of structurally reinforced concrete, rapid improvements in elevator technology, and the rise of the automobile (with all of its associated infrastructures). Atomic power and its "cleanliness" were also seen as part of the dividend of modern scientific rationality. Efficiency was a key concept: modernist designers believed that they could do more with less than the wasteful designers of the past, and in doing so build quality housing at a much lower cost per unit in high-rise buildings with a view for every resident.

The stark aesthetic of modern architecture was always somewhat polarizing. By the mid-1960s, modernist urbanism was seen less as an inevitable transformation than one side in a conflict between it and a more traditional, humanistic set of values. Many of the vaunted high-rise residential towers built for people of modest incomes in the aftermath of World War II became symbols of failure: poorly maintained, ugly (in the eyes of many), and associated with poverty and crime. Some of these towers still stand, but over time many have been razed rather than adequately maintained (Aoki 1993).

Modernist urban advocates argue that the problem with these buildings was not theory but execution: they were built at too low a price and without sufficient quality control because they were for poor people. These advocates point to modernist office buildings, built in the same era, which, with ongoing renovation and revision, have stood the test of time. In these circles, the Barbican Estate stands as one of the best counter-arguments to the idea that mid-century modern urban theory was misguided. With sufficient resources, they argue, the modernist city could have been realized.

See also related boxes "Ecomodernism" in chapter 5; and "Asian Futurism, Sinofuturism, and Orientalism" in chapter 6.

Interlocking levels of walkways and amenities at the Barbican Estate[4]

when removed, show the grain and pattern of the wood in their surface. In contrast, exterior concrete at the Barbican Estate was handpick-hammered to reveal the larger pieces of aggregate in the concrete mix, giving a depth of texture that also resists rain streaking. Interior concrete surfaces (in stairways and halls) were bush-hammered, a similar but more subtle effect that masks imperfections. This subtle finishing throughout required expert craftsmanship and likely doubled the cost of concrete works on site. Similar attention to detail is shown in both the selection of materials and execution throughout the estate.

While the UK government implemented reforms in the 1980s that made it possible for many people across the United Kingdom to apply to buy their publicly funded flats, the City's plans for the Barbican Estate included a mix of rentals and units for purchase from the early days. The complex includes a mix of apartment sizes, from small studios up to units for families.

In this way, while the estate is not very mixed in terms of class status, it is very heterogeneous in terms of residents' progress through the life course. Relatively young rising managers in the City's major firms can find a foothold close to their offices. As they progress in life, there are options for newlyweds, families with children, and then downsizing when children leave the nest.

Today, the Barbican Estate is seen as slightly quirky yet desirable for a certain kind of upscale urban resident. The high modernist lines of the estate are just enough out of the mainstream that a Barbican apartment serves as a kind of lifestyle statement piece, like a very nice Eames chair or an Arco floor lamp (Wilson 2004).[5] In the UK context, where traditional homes for the wealthy can be centuries old, the modern lines of the Barbican Estate demonstrate an aesthetic and technological alignment with the post-World War II period without having to engage in the contemporary urban housing politics of mixed incomes or the controversies of postmodern architecture.

The estate is a site with some mix of uses. The Barbican Center, the last major piece of the estate to be completed in 1982, is a major performing arts venue in London. It is the home of the Royal Shakespeare Company, two major symphony orchestras, an art-house cinema, and a public arts library. On site there are small shops in segregated spaces: for example, a dry cleaner and a small convenience store. But the Barbican Estate's relationship with its surrounding urban matrix is more permeable from within than without. The City is for its residents, even if the Barbican itself is only for a select group of those who can afford its perches.

Walled against casual passersby who are not residents of the City, the Barbican is a high modernist machine for living in the sky (see Le Corbusier 1927). Aesthetically, it evokes the early days of the jet age and a British hope that its postcolonial era would be politically and economically ascendant. Yet this residential enclave, buffered by social engineering from the daily flows of the City, maintains an engineered domesticity. It is not a site of economic production but of social and household reproduction. In this sense, while the Barbican is cosmopolitan, it is more a cosmopolitanism of consumption – of enjoying urban diversity as a pleasure – than one that challenges any well-established order of everyday life. Founded by the City of London, built by the City to house enough of its executives to justify its continued political autonomy, the place frame of Barbican Estate residents stands a degree apart from the ideas that animate the City more broadly.

Where these frames align, and where they diverge

These two urban place frames share a context in the United Kingdom and (to a lesser degree than before Brexit) Europe. They overlap in space. They both see Britain, London, and the City within Greater London as sites of wealth and privilege. Furthermore, the Barbican Estate was funded and created by the City of London. The City Corporation saw a political and practical need for a residential enclave within the square mile of its borders, and built it.

Urban Cultural Geographies

Cultural geography is principally defined by understanding social expression in particular places, such as prevailing language, religion, or artistic expressions (such as music, art, or architectural styles) of an area. Scholars interested in urban culture tend to examine how different social groups express their values in the urban landscape, or how urban environments shape individual or group place identities (Arreola 1988; Ehrkamp 2005; Hume 2015; Main and Sandoval 2015). For example, geographer Daniel Arreola (1988) showed how yards of Mexican-American households in the US Southwest (such as Arizona and California) have distinct fences that delineate boundaries but do not offer privacy from the street – the boundary making being the culturally significant purpose, not social privacy per se. Some urban cultural scholarship specifically highlights the experiences of urban migrants and explores the interplay between local urban policy practice and the recognition of underserved populations (Collins 2012; Valentine 2013; Kaplan and Recoquillon 2014; Kaplan 2015).

A focus on culture emphasizes ideas about identity and how people express shared identities in place, and so urban cultural or social geographers seek to understand what scholars have conceptualized as "difference," including the contours of class, racial, ethnic, gender, and sexual identities in cities (Smith 2001; Amin 2002; Hoekstra 2018; Hunter and Robinson 2018). A significant theme within some of this work is to portray cities as sites of cultural expression for people who have not been included in dominant political power structures.

Urban art such as graffiti – also known as street art – perhaps best illustrates both the expression of identity and the conflict between marginal groups and the dominant culture. On one hand, graffiti offers glimpses into informal culture and group identity, while at the same time it can be seen as evidence of disorder and disarray. Scholarship on graffiti reflects this divide, with some research taking graffiti

Yet the idea of the City as a powerhouse of finance is distinct from the idea of the City as a site of cosmopolitan domesticity, and this is clearest in two aspects of each respective frame: (1) who counts as a legitimate denizen; and (2) the relationship with an imagined outside beyond the City. We take these in turn.

First, the by and for capital place frame emphasizes legitimate participants as those who facilitate the reproduction of capital. One way you see this is that, unlike in other parts of the United Kingdom, the franchise to vote is distributed primarily to representatives of the City's great corporations in proportion to their size. As a result, human residents constitute only roughly a third of all eligible voters in local elections, where corporate actors accountable to their employers constitute a two-thirds majority. While there is inevitably cooperation and coordination between the City and Greater London on matters like regional transportation, policing, and infrastructure, Greater London's government is ultimately accountable to the votes of its residents, not its businesses. The City is unusual in this way.

The modernist vanguard place frame emphasizes the City as a place for living. Its residents agitate for more resources and political representation, certainly. But they also jockey for different kinds of resources. They seek access to grocery stores and recreational amenities, not just to restaurants for business lunches or to faster commutes. On the other hand, the Barbican Estate's physical layout and back-to-the-street orientation contribute to an idea that residents are the true stewards and beneficiaries of the City. Non-residents are viewed in this frame as transients. In the case of workers, they benefit from the City without constituting its legitimate polity. Visitors who are not employed in the City – as in the case of concert attendees at the Barbican Centre, for example – are viewed warily as potential disrupters and interlopers, technically welcome (and economically useful) but always poised to interrupt the central character and true purpose of the estate.

Second, while both frames are internationalist in character, the by and for capital frame emphasizes international

> or other street art seriously as an art form and manifestation of identity (Pinder 2008; Iveson 2013), while others explore its suppression, removal, and anti-establishment dynamic (Smith 2015; Lee and Han 2020; Denis and Pontille 2021). Musical expression, too, represents culture that can transform space through experience, fostering alternative imaginaries of urban life (Sites 2012).
>
> These diverse works and perspectives highlight the broad array of themes in urban cultural geography. This work overall seeks to understand how people and groups express themselves in urban areas and how these expressions help to shape cities. To some degree, urban cultural geography seeks to understand what Henri Lefebvre (1996) called the "oeuvre" (or artistic work) of urban life.
>
> *See also related boxes "Freedom and Diversity in Cities" in chapter 3; "Black Geographies" in chapter 5; and "Landscapes and Power" in chapter 7.*

relationships in terms of the City's capacity to coordinate or control international activity – especially its economic dimension. As a result, while many corporate actors engage extensively around the world in financial enterprise, the City is competitive with, and jealous of, the successes of other global cities (Amsterdam, New York, Tokyo, etc.).

In contrast, the modernist vanguard frame is more oriented toward place consumption than place competition. Paris or New York provide opportunities for experiences (and sources of companionship) in a network of elite urban places around the world. Because Barbican residents emphasize London as a place of consumption, they see the City as more fundamentally integrated into Greater London than those who center the Corporation and its corporations.

What is excluded from both frames is, to a large extent, not just abject poverty but in large part working-classness in general. The median income in the City is nearly twice that of Greater London as a whole, and nearly 50 percent higher than the second-highest neighborhood in the region (ONS

2021). Poverty is for other people, not for residents of the Barbican, nor for those employed in the financial sector.

Conclusion

The distinction between the idea of the City as a place for wealthy and domestic cosmopolitanism on one hand versus a place for wealthy and internationalist financial braggadocio on the other may seem minor. After all, both frames center a kind of lifestyle utterly inaccessible to most residents of Greater London, the United Kingdom, or the world. And yet these two visions of London suggest very different modes of urban inhabitance.

Henri Lefebvre, in his classic essay "The Right to the City" (1996), rebukes the city of the mid-twentieth century as one that was built by the working class to the benefit of capital. He argues that despite their extensive power over the city, the wealthy have not actually occupied it. Instead, they have abandoned the pretense of genuinely urban life, one where rich possibility comes from propinquity, mixing, and exposure. Instead, he argues, the city has been left perforated, empty of authentic use, yet still excluding the working people and the poor. In Lefebvre's vision, the titans have retreated from urban life: they lock themselves away in suburban compounds, gated communities, or castles in the sky. They have, in this view, taken urban life away from workers and yet refused to actually inhabit it themselves.

The *by and for capital* place frame embodies Lefebvre's complaint. The purpose of the City, in this frame, is not only alienated from the working poor; it is even alienated from wealthy residents, allowing them token votes of resistance in a political system designed to preserve the power of corporate actors. If the City of London is for capital, by corollary it is not for living.

But the *modernist vanguard* frame complicates this story of the City. This place frame still excludes, true; but it does not leave the City's landscape empty of occupation. Instead, it articulates a version of occupation that shows the idealism of wealth. It is richly used, filling spaces designed to be plurally inhabited. It is a version of London that absents the poor; but it does not absent urban life.

As is always the case, there are other frames of the City. The governments of Greater London and the United Kingdom often downplay the separateness and specialness of this ancient enclave, maneuvering slowly over centuries to one day revoke its special status. Those other imaginations frame the square mile as less exceptional and also less apart. Many from outside the United Kingdom are oblivious to the distinction between the City and the city; they only see London, obliterating any distinction. But for those who live and those who work within the line of its ancient walls, the difference in purpose can be profound.

This chapter might help you to think about cities like New York, United States; Singapore; Hong Kong; Taipei/New Taipei; and the Vatican City/Rome.

–3–
Tehran: Islamic Developmentalism/Diverse Cosmopolitanism

Azadeh Hadizadeh Esfahani and Deborah G. Martin

Tehran is the capital of the Islamic Republic of Iran. It is a so-called "Global South" city, capital of a country that is not considered a major player in, or driver of, the global economy. Yet it is nonetheless a part of the global economy and embodies some of the classic challenges of cities in so-called "developing" economies, including concentrated urban power and wealth, rapid economic change, significant economic inequality, and considerable environmental degradation. It is a major city in an oil-rich nation, and it exercises some global power. (That power is mediated by its country's global political standing – for example, intermittent financial and other sanctions against its economy over its uranium-enrichment program throughout the 2010s and into the 2020s.) Additionally, as the capital of an explicitly religious nation-state, Tehran juxtaposes a unitary theistic governance framework with a population diversity that is typical of urbanized areas worldwide. It thus offers a plethora of competing meanings and forces to consider.

Tehran is a large city and has experienced significant growth in the past three decades, a trend similar to many other cities as urbanization proceeds rapidly around the world and especially in the Global South. According to the census of 2016, Tehran had more than 8,500,000 inhabitants who are spread in an area about 730 sq. km, which makes Tehran among the most populated and dense cities of the world. Together with the towns and smaller cities around it, metropolitan Tehran is home to more than 14 million people,[1] most of whom commute daily in or to the city of Tehran. The city's population is partly the result of high birth rates, but more importantly also to migration from other urban and rural areas of Iran (and even adjacent neighboring countries) of people seeking access to the wide variety of opportunities and services concentrated in Tehran. Being the capital of the country with a very centralized government structure has made Tehran a key actor in the political, economic, administrative, social, and cultural affairs of the country.

Two of the most prominent views of Tehran are linked to the idea that it dominates the economy and politics of Iran, framing the city as a place of *economic intensity and growth* and also as a *cosmopolitan site of diverse modes of life*. These frames play off of each other, both drawing to some degree on a trajectory of development in Tehran and Iran as a whole. Focusing first on the frame of economic growth, Tehran is a place of development and a signifier for the economy of Iran as a whole.

City of economic intensity and growth

It makes sense to think about cities as engines of economic development. Developmental narratives about cities start with an axiomatic assumption that growth (in dimensions such as economic output, population, and/or comparative cultural importance) is a positive, necessary trajectory for cities to succeed economically. A key idea underlying belief in "development" is that over time successful cities, regions, and even civilizations go through a process of improvement and intensification called "development." The idea of development is metaphorically related to evolutionary processes in organisms or ecosystems where systems adapt over time and become more capable. A good city, in developmentalist narratives, becomes a more ideal form of itself over time. The path may wind, but growth is a formal measure of a city matching a kind of urban ideal. For a city government, facilitating development means ensuring the scaffolding for ongoing growth: arranging suitable transit for people and goods, wooing a workforce with needed skills (whether that be to assemble goods or ideas), and luring other capital investments in buildings and utilities to foster a culture and expectation of continuous growth.

Proponents of developmental urban narratives tend to view growth as possible and desirable everywhere, while acknowledging that different cities (or countries) will have different forms of growth. Economists called these different forms of growth "comparative advantage," and some urban scholars emphasize that, while all cities have some degree of economic and social complexity, some will have more of some types of industry, commerce, and entertainment than others. For example, London is a global hub of financial activity, while Los Angeles has a much more influential media production industry. Some cities' potential for ongoing growth is based on providing tourist services, others on shipping extracted goods to factories elsewhere, others based on producing manufactured goods, and still others managing the systems that connect all of these disparate economic elements.

Developmental urban narratives are often critiqued by academics because they have historically been associated with politically and ecologically problematic ideologies. The idea of "development" has for several centuries been linked to paternalistic, racist, and colonial concepts of the cities and economies of imperial nations in the Global North as "developed," with the "less developed" or "undeveloped" cities of the Global South conceptualized as childlike. Despite these scholarly critiques, however, developmental ideas that center perpetual growth and inter-urban competition in the urban project are widely shared by government and private-sector actors. The idea that development is good and necessary goes almost unquestioned in most "real world" urban governance decision-making processes. At the same time, however, in any city, growth benefits some people (in particular in economic sectors) more than others.

In Tehran, as in Iran as a whole, developmental approaches to the economy have long shaped the trajectory of growth and the coming together of diverse people and infrastructures. Tehran may seem very different from other cities as a city located in the Global South, ruled by an explicitly religious government and benefiting from natural resources spread over the country (particularly oil wells and mines). The particular details make Tehran, the place, distinct; but, like other cities, it also embodies key urban characteristics and processes that help us understand cities broadly. More than 200 years ago, Tehran became the capital of Iran and it remained the capital of the country through the Pahlavi era from the 1920s to the present day. During all these years, the leaders of the country all sought to represent Tehran as a modern and developed city comparable with other global cities. We identify three main periods of development in Tehran's recent history.

Modernization of Tehran started in the 1920s when Reza Khan, founder of the Pahlavi kingdom, came to power. His approach was centralized top-down development. In his view, development could be achieved through rapid economic growth, not an incremental comprehensive process of change. During his 20-year reign, the gates and walls of Tehran (which marked the city's borders but were also used for defense) were removed and Tehran was extended in all directions. New buildings and roads were constructed. The "first" of many organizations and buildings were founded during these years: the country's first university, hospital, airport, stadium, bank, library, museum, and cinema were all built in these twenty years under Reza Khan in Tehran (and, in the years following, similar developments occurred in other Iranian cities). Industrialization, the assumed path toward (economic) development, led to the establishment of a variety of factories for the production of things like cement, military equipment, and tobacco in and around Tehran. These investments intensified the power of Tehran and added an administrative role to its political and economic role in the country. All these changes made Tehran an attractive destination for migration, and its population increased from less than 200,000 in 1920 to more than 500,000 in 1940.

The second wave of development in Tehran was the period of 1960–1970. During this period, Mohammad Reza Shah Pahlavi initiated several programs, mainly focused on improving the economic and military power of Iran. His desire to represent Tehran as a modern developed city led to significant changes in the city. Tehran's first master plan was developed during this period: Tehran was divided into ten regions, with specific plans for expanded infrastructure for industry and housing in each. Streets were widened and new highways were built. During these years, five new universities were founded in Tehran to educate the required labor force and accelerate development of the country. These developments were possible due to the wealth coming from oil exports, as well as the good relations Iran held with many Euro-American countries at that time, which enabled it to leverage international financial support for growth. By 1970, Tehran had more than four million inhabitants (by comparison, both London and New York City had over 7.5 million people, while Tokyo had over 20 million). The changes in Tehran, and its central role in the economy of the country, served as a draw for people across the country. Many of these new arrivals, however, lived in informal settlements at the margins of the city because the rapid growth was spatially and economically uneven.

The years of 1975 to 1990 were years of protest, revolution (which in 1979 changed the government to an Islamic Republic), and war. (The war, which lasted from 1980 to 1988, was driven by Iraq's worries that the new Shia-Islamic state in Iran would destabilize Iraq's Sunni-led government.) After the war, Tehran's renewal became a key focus for the national government, and its population grew from about 6,500,000 people in 1990 to almost 9,500,000 in 2021. Over the same period, the average price of a house increased

The modern landscape of Tehran[2]

577 times, while the minimum wage increased 333 times. It is perhaps ironic that Tehran, the capital of a country ruled as an Islamic state which came to power by the "revolution of downtrodden and slum dwellers,"[3] has experienced such extreme and unequal economic growth.

Despite the revolution and its anti-western slogans and critiques of modernization in the style of the United States and Europe, the 1980s saw a renewed focus on mega-projects: highways, tunnels, malls, large entertainment complexes, and symbolic and tourist-oriented projects. Operationalization of these mega-projects and other development plans required vast amounts of capital (much of which was generated in the oil economy), which in tandem with privatization of land and industries created a new group of powerful actors involved in the growth coalition (see box, "The Growth Coalition") who were focused on increasing the value of real estate in Tehran.

Growth-oriented actors in Tehran can be categorized in three groups. The first group consists of developers who build mega-projects. These developers have significant capital and land resources, as well as personal connections with government officials. They are actively involved in shaping the development narrative for Tehran, and they gain huge profits from the rearrangement of space (for example by changing from residential land use to commercial, changing the route of a highway to increase access for areas where they own land, or building large projects in neighborhoods that question the need for them). This small group of large developers is powerful and influential. Its power has roots in ongoing collaboration with security organizations, military installations, banks, and huge semi-public foundations that provide services to underserved populations on behalf of the government.

The second set of actors are not directly involved in mega-projects but are able to invest in nearby properties or businesses in order to reap profit from the development activity. These local builders and investors are able to access development plans via their relationships with municipal staff, government officials, or the media. These actors use

The Growth Coalition

Urban growth requires more than simply population increase; it is also fundamentally a process of real-estate development so that housing, transportation networks, and business development occur in ways that encourage greater population and economic activity in an area. Cities *take up space*; they grow, people interact, move around, and exchange goods and ideas with each other in places that are made on actual ground. Referencing this reality, Harvey Molotch (1976: 309) argued that "land [. . .] is a market commodity providing wealth and power, and [. . .] some very important people consequently take a keen interest in it." He called these influential people the "growth coalition"; this concept was articulated specifically in relation to US cities but has resonance in other contexts. Fundamentally, the growth coalition refers to a combination of urban denizens – leaders, residents, business owners, transit companies, etc. – who are continuously interested in, and work to participate in, land development. Many urban actors benefit from overall growth in the city, even if they don't directly invest in it and profit from it. These actors include local politicians, property owners, construction and development businesses and workers, and real-estate agents. Even local media companies participate in the growth coalition; they rely on a local business base to purchase advertising and a local population base to subscribe to their services. Growth, to them, means more customers. The many members of the growth coalition, then, share the perspective that urban growth is a benefit to the city; land investment and renewal is inevitably good for the city's overall economic status – even when these changes cause significant displacement and negative impacts for some residents and (often small) businesses (Wang 2020; Müller, Murray, and Blázquez-Salom 2021).

Research that examines how growth coalitions function and their impacts on urban development and social life in cities spans a range of contexts. For example, Jeff Broadbent their positions, knowledge, or relationships to find a way – such as bribing, postponing the public issue of a plan or enforcement of a regulation, or simply acting first on inside knowledge – to circumvent local-level government procedures in order to proceed with their projects.

A third category of growth coalition actors are individual residents who also seek to profit from other development or land-use changes but do not have the capital or relationships to do so at a significant scale. These actors are not as powerful as the first and second set of actors, but they are able to take advantage of the changing rules and conditions at the level of city government and its enforcement of regulations. For example, on different occasions, the Tehran municipal government allowed owners to increase the number of floors in their buildings, as a way to increase land values and, thus, tax revenues.[4] This change created an incentive for building owners to demolish and rebuild their buildings, where they would not otherwise have done so, in order to increase the number of floors and gain profits from increased rent. In some cases, such activity could spur these actors to get more involved in the real-estate field as they invest the new profits from expansions into other buildings or new developments.

These three economic growth-oriented groups typify the theory of an urban growth coalition, interested in fostering development in their city to increase population, land uses, and land value (see box, "The Growth Coalition"). These development and finance-connected actors work with other actors, like real-estate dealers, notary and registry staff, municipal staff, justice department staff, and the media, to make this coalition work and foster growth in Tehran's real estate. While the details of how these actors function are particular to Tehran's specific mores, their existence is typical in any city. For this system of real estate to realize added profits, they need continued and diverse demand for housing and business activity. Tehran's role as the capital of Iran, and a site of ongoing development, fosters such growth. Yet this growth is uneven geographically within the city and in relation to the country as a whole.

> (1989) argued that business leaders assumed tremendous importance at both local and national scales in a development project in Oita Prefecture, Japan, over and above the strong state context. In China, however, Zhang (2002) argues that a strong local state is dominant in the growth coalition, which focuses primarily on economic dimensions of cooperation and less so on political coordination. Although growth has been presumed to be the dominant focus of business, state, and society coalitions in cities, some scholars find that community advocacy can help support a coalition to move away from solely business-led growth in order to foster community cultural land uses (Joo and Hoon Park 2017). Growth coalitions are shaped by local and national contextual forces, then, even if economic growth and land investment principles persist globally.
>
> *See also related boxes "Urban Economic Processes" in chapter 2; "Municipal Development and Finance Strategies" in chapter 4; and "City Branding" in chapter 6.*

Unevenness of growth – Tehran as a swallower of resources

For many people in Iran who worry about overpopulation and stretched national capacities, including development practitioners, public officials, Tehran residents, and activists of less developed cities and provinces, Tehran absorbs too much natural, economic, and human resources from all around the country (see box, "Urban Agglomeration"). In this framing, which offers a negative view of the developmentalist orientation, Tehran is a city which swallows the resources of the country.

Land value in Tehran is relatively high compared to other Iranian cities (average land value in Tehran is about twice the average land value in Iran) but still a significant gap exists between land value in the wealthy north and that in marginal southern parts of the city. As with many cities, Tehran's landscape includes symbols of both rich and poor: mansions, expensive cars, and luxury goods in northern Tehran, as well as shacks and slums mostly in the southern parts of the city.

Tehran's share of the natural, economic, cultural, and human resources of the country is disproportionate to its size and population. The city takes up about 730 km^2, or less than 0.05 percent of Iran's area, and hosts more than 10 percent of Iran's population. Tehran has been criticized by many development practitioners and governmental officials from other provinces who argue for more equal access to and distribution of resources. To these groups, Tehran is a city which absorbs resources, talents, and capital from all over Iran, consumes them with no (or low) return to them, and increases the gap between itself and the rest of the country.

Tehran itself does not have any significant natural resources (no mines, oil wells, rivers, or forests); but it benefits from the natural resources all around Iran even more than the cities and regions that host these resources. Multiple rivers (like Karaj, Jajrud, and Taleqan) in the Alborz Mountains north of the city pass through towns and villages around Tehran. For years, these rivers supplied underground water, aqueducts, flood canals, and brooks going along the streets and avenues of Tehran and addressed Tehran's water demands. But as Tehran grew, the traditional methods became insufficient, and in recent decades a system of dams has become the main source of Tehran's water supply. Thus a number of regional dams have been constructed to provide water resources needed for Tehran. Additionally, Tehran residents also receive more subsidies for water costs than those in other Iranian cities. Similar patterns exist in the consumption of other utilities like electricity and gas. Tehran's share of the total gas and electricity consumed in the country is more than the average per capita of the country as a whole. However, all these utilities receive subsidies from the government and are used by people at a cost much less than their production cost. Therefore, Tehran receives a higher proportion of subsidies and financial resources compared to other cities.

The economic imbalances between north and south within Tehran, and between Tehran and the rest of the country, are

Urban Agglomeration

Geographers, economists, and others have pointed to the significance of large geographically contiguous urbanized areas within regions; as Fang and Yu (2017) note, a number of terms have been used worldwide, since the early twentieth century, to characterize these spaces: conurbation, metropolitan area, megalopolis, urban cluster, city-region, and agglomeration, among others. These words all connote geographies with relatively densely settled and built-up, economically integrated urbanized areas. Typically, an urban agglomeration has one or more central cities, in hierarchical relationship with many surrounding urban or suburban areas that have a high level of socio-economic interconnections (Gottman 1957; Fang and Yu 2017). The term "agglomeration" characterizes a spatial form, but one that has evolved through processes of economic and physical growth in a region. Central to the idea of agglomeration is the presence of several cities and towns within a geographical region that are linked by transportation, information, and other socio-economic networks (Gu, Chai, and Cai 1999; Fang and Yu 2017). The spatial form is of interest to scholars because of the integrated economic activity that occurs there, spurring urban growth and blurring the lines between municipalities in terms of urban planning, environmental impacts, and social life (Teaford 2006; Fang 2015).

Understanding urban agglomeration as a spatial form emphasizes the infrastructure and interconnections of urban regions, including roadways, rail lines, commuting patterns, shipping routes, electricity, and internet access and storage points. Agglomeration can also be understood as a set of social and economic processes by which cities and towns become interlinked, through commuting, supply-chain networks, labor markets, etc. Thus agglomeration as a concept draws attention to how regional transit systems and economic dynamics shift over time.

Some municipalities within an urban agglomeration try to entice companies to their jurisdictions through strategies not new and existed before the Islamic Revolution of 1979. But after years of an Islamic state ruling the country which emphasizes bringing justice for all, removing corruption, and improving quality of life, especially for economically deprived groups, the imbalances within Tehran and with the rest of the country have increased. The population size, economic might, and political centrality of Tehran cause many resources throughout the country to be diverted to Tehran. The wealth produced from oil and gas resources in the southern provinces of Iran, from industrial complexes all around Iran, and from many other resources all come first to Tehran and then are distributed to other cities. Banks in Tehran hold about half of the total amount in savings accounts in Iran, even though the population of Tehran is just about 10 percent of Iran's population. The average Tehrani resident receives about three times more interest income than the average Iranian. The average income of an Iranian family was 540 million rials in 2020, while the average income of a Tehrani family was 810 million rials.

Most business loans and financial facilities are concentrated in Tehran, so many businesses have offices there to have better access to (especially financial) facilities and opportunities. Eventually, business representatives who regularly travel to Tehran for business decide to purchase property there or to move there. Migration to Tehran is widespread across income levels as people seek to access Tehran's opportunities and resources and its diverse pool of jobs. According to the Iranian census in 2016, 20 percent of the total internal migration was to Tehran, and 88 percent of the increased population of Tehran from 2011 to 2016 was due to migration.

The concentration of political, economic, cultural, and human capital in Tehran has created an assumption that whoever is looking for growth, progress, success, and achievement should come to Tehran. Many teenagers who choose Tehran for continuing their studies decide to stay in Tehran and not return to their home cities because of Tehran's job opportunities, cultural diversity, and diversified social relations. A broader media and popular discourse

> such as taxing, labor force training, infrastructural investments, and so on. A persistent tension within urban regions is the political differentiation of municipalities within the broader agglomeration; some regions address these tensions with coordinated regional land-use planning to address economic and social inequalities within the region, as well as the overall growth pattern, transportation infrastructure, and environmental impacts (Gleeson and Spiller 2012; Zimmerman, Galland, and Harrison 2020).
> See also related boxes "Urban Economic Processes" in chapter 2; and "Sprawl, Density, and Urban Growth Boundaries" in chapter 5.

portrays Tehran as a destination for living for diverse groups of people. Tehran, in this narrative, is a *cosmopolitan site of diverse modes of life.*

Tehran as a city of choice

The broad picture of Tehran in the developmentalist frame emphasizes its economic and political centrality, as well as individuals, whether they be wealthy developers or small-scale landowners, who can leverage their capital and connections to contribute to the city's emphasis on infrastructure and land development for economic intensity. For many city residents, however, the broad macro-scale measures of regional economic growth or the footprint of the city hold less meaning, except perhaps as part of civic pride and identity, or highlighting the economic gaps between people who manage land development and those who just struggle to live. The forms of growth that come with development often change a city in substantive ways, including removing old buildings, reconfiguring streets, and raising the costs of rents in nearby housing. Residents' experiences of urban places are often bound up in the minutiae of day-to-day patterns and the rich experiential or representational character of small interactions. The smells of unfamiliar cuisines in an apartment block, the range of languages spoken in the playground at a park, the diversity of ways that people dress and interact – these everyday experiences of a city may be shaped by higher-level economic decisions, but they also offer a different set of understandings of urban life.

An important way that people experience the city is recognizing differences between their own lives and those of other people. In the developmental frame, these differences are broad and often international; before the establishment of the Islamic state in 1979, Tehran's leaders broadly followed the development models espoused by European and North American nations, and over the decades increasingly gestured at importing socio-cultural norms from those countries as well.

The revolution was intended to usher in a utopian Islamic society, and Tehran was articulated as a key site where these ideals would be realized. These ideals have not always been realized in present-day Tehran. In theory, there are specific ideologies ruling Tehran (and Iran) in accordance with religious justice-oriented and anti-capitalist norms; but in practice many of these norms have been stretched and have aligned themselves with urban strategies elsewhere in the world.[5] This flexibility and compliance affirms that being in a globalized world necessitates a degree of openness to diversity and difference. National openness to adaptation to the realities and conditions of the global economy also shapes daily life in Tehran.

Like any large city, Tehran offers its residents or visitors a large degree of anonymity and relative freedom. Even within an Islamic framework, living in Tehran provides people a degree of social freedom from family and traditional relations. Compared to smaller Iranian cities and towns, where one may occasionally encounter families or friends in the street and thus feel pressure to behave in accordance with specific norms (especially religion-related norms in the case of Iran), in Tehran one is unlikely to see a familiar face in its busy and densely populated streets. This freedom and anonymity is a common urban characteristic and not particular to

Tehran; indeed, it is a hallmark of how urban scholars over a hundred years ago theorized the significance of urbanity (see "Freedom and Diversity in Cities" below). Yet in a formally conservative religious state like Iran's, the cultural, environmental, infrastructural, and social characteristics of Tehran provide its residents with a degree of acceptance of difference and tolerance in dealing with divergence from formal legal norms that is distinct from any other place in the country.

Women's clothing represents a particularly illustrative case of Tehran as a city of choice. The formal rules of the country obligate women to cover their hair and wear (long-sleeved) clothes which cover their bodies to their ankles and wrists. Therefore, the formal clothing of women is either a chador[6] or mantoo[7] scarf. Walking in the streets of Tehran, women do wear clothing that covers almost their whole bodies, but there is a wide range of clothing among them (see image below). On one hand, there are women who wear black chadors with only their eyes and nose visible, without makeup and sometimes even with gloves. On the other hand, there are women who have a piece of material on their heads with hair coming out of it both at the front and back. They wear a kind of mantoo, but their mantoo is a colorful thin cloth that permits others to see the clothes, and the shape of the woman's body, under the fabric of the mantoo. These mantoos have no buttons, clasps, and so on and are open down the front. They are usually short, falling at or below the hip with sleeves from shoulder to elbow. Often these women also wear full makeup, including polished nails visible in sandals. While conservative norms would dictate that women's hair be completely enclosed, in casual settings the headscarf would often be loose and falling back, with much of the hair beneath visible. Between the extremes of the fully covering chador and the sheer mantoo, women choose a spectrum of clothing options. Although the short, sheer mantoo and visible makeup are not formally accepted by the government, it is informally accepted and tolerated in Tehran, whereas in many other Iranian cities there are socio-cultural pressures preventing women from wearing this type of clothing. Riding bicycles in the streets, playing musical instruments in public, being a waitress or taxi driver are other examples of behavior accepted for women in Tehran but typically not in other Iranian cities. Although women's clothing and behaviors may be the most observable example of Tehran's tolerant culture, there are others as well. For example, the cultural environment of Tehran is more open to non-traditional relations, especially friendships between men and women or even so-called white marriage, which is a non-religiously sanctioned sexual relationship.[8]

This openness to social diversity is evident in other aspects of city life. Individuals with different ethnicities from different regions of Iran have migrated to Tehran looking for employment opportunities, sociocultural openness, better access to services, and so on. Therefore, in Tehran it is not unusual to hear people speaking languages other than Farsi (which is the formal language of the country), such as Turkish and Kurdish. In addition, most of the foreign embassies, central offices of multinational companies, and head offices of companies with foreign relations are located in Tehran. Thus people from other nationalities can be seen in Tehran and languages other than Farsi, like English, French, Arabic, and Chinese, can be heard in the streets. Tehran is not a border city but nonetheless hosts a considerable Afghani population in addition to an Armenian population that has been living in Tehran since 1940.[9] As a consequence, in addition to several mosques, Tehran has churches and synagogues, as well as Chinese, Mexican, Afghani, Greek, Italian, Thai, and other restaurants. Other types of infrastructure, such as a wide range of public and private universities and schools, sex-segregated sports clubs, theaters, and music venues, are prevalent. This access to diverse options is matched by a choice of diverse lifestyles.

For residents who have the financial and social means to emphasize choice as a central experience of Tehran life, even the climate and natural environment provide a range of options. The north of Tehran is mountainous, whereas the southern parts of the city reach a desert. When heavy snow covers mountains, one can go skiing or enjoy the beautiful sky of the desert. When it is too hot in the desert,

Freedom and Diversity in Cities

The notion that cities offer social relations that are freeing from the bonds and expectations of kinship is an old one (Sorokin 2002; Tönnies 1995 [1887]). Louis Wirth (1938) argued that urban residents experience freedom from close personal ties and related expectations for behavior, and that urban areas have a greater diversity and range of human behavior. Al-Haj (1988) argues that in societies with strong patrilineal kinship arrangements, urbanization can foster greater focus on nuclear families and opportunities for women outside the home.

Certainly a wide range of scholarship points to a freeing of societal norms and a greater range of social expression in cities, from opportunities for a broader range of sexual expression and identity (Wilson 1991; Chauncey 2008 [1994]) to a greater willingness to engage in political activism (Schoene 2017). Undergirding these ideas of freedom is the theme of diversity; it is the coming together of people of different backgrounds and cultural norms that enables urban freedoms because it makes for different ways of being explicit. At the same time, different modes of life, the cacophony of city streets, can be disorienting and distressing (Wilson 1991). Yet it is both the exhilaration and fear of difference that, some scholars argue, can be generative of the creativity that spurs urban economic activity and growth (Florida 2002, 2014). Diversity is, therefore, a critical feature of urbanism (Wirth 1938).

The tensions between economic forces and artistic expression are quite explicit in some urban theory; Henri Lefebvre (1996) argued that it is everyday working people who create the *oeuvre* (or "work of art") of the city, in contradistinction to the powers of corporations, urban planners, and officials to make urban landscapes. Smith (2005 [1996]) takes this idea further, arguing that artists who creatively reuse disinvested urban spaces such as warehouses for work–live studios end up benefiting urban real-estate interests by helping to create new markets for urban investment

Examples of women's bodily comportment and dress in Tehran[10]

> and renewal. Sharon Zukin (1989, 2009) recognizes the potential of co-optation of cultural expression for commodification but ultimately argues that authentic urban spaces can be sites for community expression, rather than merely investment opportunities (Zukin 2009). In this vein, Wright and Herman (2018) argue that artists and residents of neighborhoods struggling against outside investment and gentrification pressures can create temporary articulations of community power and identity through street parties and performances. They suggest that these temporary "alternative spatial projects" are especially important for economically vulnerable and marginal populations, such as Black communities in the United States, to claim urban places as their own. Urban spaces are thus sites for free expression by artists, community groups, and others, while simultaneously holding potential for constraints on that expression by politically and economically powerful actors.
>
> *See also related boxes "Urban Cultural Geographies" in chapter 2; "Black Geographies" in chapter 5; and "Landscapes and Power" in chapter 7.*

one can enjoy the cool weather of mountains. The location and geography of Tehran makes it a four-season city which allows a great variety of activities (agricultural, sport, or recreational). It does not have a forest, but it hosts a few forest parks. No significant river passes through Tehran, but rivers are accessible within an hour's drive from any location in Tehran. While these elements are not equally accessible to all residents, they nonetheless form part of a place frame of choice.

Frame synergies

The wide degree of choice and openness to diversity and difference in Tehran is certainly related to being the country's capital and having a central role in the national economy and politics. But elites and growth coalitions play an important role in maintaining and amplifying Tehran as a city of choice, thus connecting the frames of Tehran as a site of both *developmental economic intensity* and *cosmopolitanism*. One of the supporting justifications of the growth coalition shaped around real estate, for example, is to celebrate Tehran's multiculturalism and present Tehran as a city for different groups of people with different lifestyles, attracting newcomers, and thus justifying multiple development projects that also enable elites to reach their rent-seeking goals.

At the same time, actors involved in the real-estate growth coalition are not the only group who frame Tehran as a city of choice. Another significant group picturing Tehran as a city of choice are the elites of the smaller cities and towns who have migrated (or want to migrate) to Tehran for different reasons; for them, choices in Tehran are aspirational. Throughout the country, students seeking to go to university must take national entrance exams, and those who do well often choose to attend Tehran universities, in part because they are the most prestigious. These students typically then want to stay in Tehran as the city offering the most choice and opportunity in employment and lifestyles. The imaginary of Tehran as a city open to diversity and choice is meaningful to many constituencies, who perpetuate the characterization and argue for continued individual freedom in the city. These groups include, for example, people who can afford to maximize choices, such as wealthy families whose kids can go to a horse-riding complex; religious minorities who seek freedom from Islamic behavioral expectations; or business people whose work depends on exchange and relations with foreign countries.

The reputation for flexibility and openness in Tehran is also itself a tension. For example, some people protest to the government, advocating more social controls on individual behavior in relation to clothing or public non-Muslim religious and cultural expression. Understanding Tehran as a place of difference where multiple religions, as well as a range of Muslim religious expression, are tolerated as part of urban life means that some people in other parts of the country reject its permissive culture and stay away from it. Some people living in smaller cities of Iran represent Tehran negatively as a place they do not want their kids (especially daughters) to go for study or work. This rejection of the symbolism of Tehran as permissive extends to other framings of the city as a negative for the country as a whole.

Seeing Tehran as a cosmopolitan city or a city for growth and success has attracted many people to Tehran. But the appeal of the city does not mean that the quality of life in Tehran is either acceptable or pleasant for all its residents. The average land value in district 1 (which is located in the north of Tehran) is 5.5 times higher than the average land value in district 18 (which is located in the south of Tehran). Indeed, the concentration of resources in Tehran does not mean that Tehran itself is a balanced city. The gap between the wealthy population of Tehran living mainly in northern parts of the city and the poor population living mainly in the marginal southern parts of the city is greater than in any other city in Iran. Of the water consumed in Tehran, 47 percent is consumed by three northern districts which host about 17 percent of Tehran's residents.

The municipality of Tehran has, in the last two decades, sought to bring more opportunity for participation and agency in the growth of the city to its ordinary, non-elite residents. In doing so, the municipality draws on the frame of openness and choice, inviting residents to engage in local policy decisions in their neighborhoods. At the same time, however, these engagements directly challenge some aspects of the municipality's developmentalist frame, as some residents have mobilized to protest mega-development initiatives in their areas, such as new roadways or hospital expansions.

While the developmentalist place frame highlights the structural forces of government and business relations and interactions at a large scale, these neighborhood-level organizations highlight the individual and collective agency and everyday interactions of residents at small scales. The more open cosmopolitan frame and environment of Tehran (in comparison to other Iranian cities) has played a supportive role in the mobilization of activism of residents around topics such as preservation of green space, opposition to mega-project development, establishment of participatory neighborhood/urban management systems, etc. Although in many cases mobilizations may have not reached the desired goal (in terms of preventing a development project or construction in a green space, etc.), they have nonetheless influenced the social atmosphere of some neighborhoods in terms of bringing residents together and giving them confidence for making political claims against developments or in favor of green space provision and community spaces.

Conclusion: overlapping frames and hints of other imaginaries

As described in the Introduction to this book, urban place frames reveal and emphasize different properties of cities, including specific actors, relationships between people and governments, and meanings. In Tehran, the pervasiveness of the developmental frame for explaining the city's growth orientation throughout the twentieth century can, on the one hand, demonstrate the persistence and effectiveness of economic explanations of urbanness. The economic drive for continuous development of real estate and infrastructure survived a political revolution and was incorporated into an Islamic political ethos, enhancing a growth coalition in the city, despite its seeming contradiction with core aspects of Islam. So the explanatory power of growth and development is significant. At the same time, the developmentalist frame also tends to augment, or at least does not conflict with, a frame of diversity and support for cosmopolitanism.

Cosmopolitanism, or simply lifestyle choice, also situates Tehran for its residents as a site for openness in daily life. This cosmopolitan frame is compatible with developmentalism, but the economic modes and cultural modes do not, in fact, require each other. They coexist, somewhat uneasily. Indeed, the greater range of lifestyle choices enabled by openness also empowers residents to self-organize and promote local-scale infrastructure like green spaces and smaller building footprints, challenging the growth imperative. This activism offers some explicit challenges to the developmental frame. The frames explored here offer hints to the ways that some highly promoted and quite explicit place frames also create opportunities for other perspectives to become more visible. More exploration of frames in Tehran would examine the types of claims evidenced in neighborhood organizing by some Tehranis, and how they intersect with other understandings of the city. Recognizing that place frames can contradict and yet also shape one another points to possibilities for political and social coalitions around alternatives.

This chapter might help you to think about cities like Istanbul, Turkey; Cairo, Egypt; Pyongyang, North Korea; and Lagos, Nigeria.

–4–
Worcester: Local Economic Engine/Regional Forest Under Threat

Worcester, Massachusetts, United States, is a city that is perhaps best known as the "second-largest city in New England," a moniker that highlights its relationship to the "first" of Boston. Boston is "first" because it is also in Massachusetts, and the capital city of that state, and is unquestionably the largest city in the New England region – typically understood as the US states of Maine, New Hampshire, Vermont, Massachusetts, Rhode Island, and Connecticut. (In 2020, US Census results showed that the city of Boston had a population of 675,647, while the city of Worcester had a population of 206,518 – both considerably larger when their suburban populations are included in metropolitan area counts.) As a "second" city, Worcester often seems buffeted by outside forces, whether these are decision makers who determine whether Worcester's aspirations as a commuter city to Boston can be met by a regional train schedule; companies that can make or break local plans for downtown growth through their decisions to locate in Worcester or someplace else; or federal agencies that dictate and coordinate policy when an invasive beetle is discovered in the city. In Worcester in recent years, all of these dynamics have been at play, highlighting the limits of local decision making for cities that are always constituent parts of broader regions, economies, and polities. Fundamentally, questions of urban growth and local policy have iterated with broader regional concerns in ways that point to the tensions between a local place identity and broader economic and social forces.

Worcester, Massachusetts, like many cities in the postindustrial "rust belt" of the United States, has struggled for at least 40 years to transition its economy from its industrial heyday in the first half of the twentieth century. The population of the city reached just over 200,000 in 1950, then experienced a steady decline until the early 1980s. By 2020, however, US Census results indicated that the city had finally rebounded to over 200,000 persons. The trajectory of Worcester's decline and nascent regrowth forms one narrative which focuses on the functions and built environment of the *downtown area as the key to economic vitality*, obscuring much of the rest of the city. In this downtown frame, the key actors are often government officials, particularly at the city level, who market the city and shape its downtown by obtaining federal and state funding support to leverage private real-estate investment. Some of this leveraging explicitly drew on notions of Worcester's pride of place. Another narrative, which emerged after 2008, situates the *city as a place for nature*; specifically, trees, in a moment of threat from an invasive insect and which played out primarily in residential neighborhoods. Key actors in this frame include residents who value the aesthetics of tree-lined streets and the privacy and shade of backyard trees, and state and federal actors who frame urban forest management as part of a regional strategy aimed at protecting New England's tourist and lumber-oriented economies. Both narratives link formal landscapes – downtown's building configuration and the natural environment – to the health and vibrancy of the city overall. Yet the form and scope of each landscape varies wildly, making some aspects of the city far more salient and meaningful than others. State actors, namely different levels of government officials, also play a significant role in both frames in what is done for and to the urban landscape, highlighting the significance of political relationships and dynamics "beyond" a city, which nonetheless shape its landscapes.

Remaking downtown, again

Starting in the late 1960s, Worcester city leaders, working with land developers, pursued a number of strategies aimed at retaining *downtown as a regional economic center*. A downtown mall opened in 1971, followed by a civic center in 1983 that for many years was a regional hub for entertainment events, especially music concerts, before being overtaken by other, newer regional alternatives closer to Boston. The mall was an initial success but then was swamped by more suburban development, a trend common for many American urban downtowns.

In the 1990s and early 2000s, Worcester sought a broader economic strategy by trying to intensify its connections to the regional economy of its larger, booming neighbor about 50 miles to the east, Boston. Commuter rail service to/from Boston returned to Worcester in 1994 (it had been cut entirely in the 1970s) and by 2000 the city had a gleaming, restored downtown Union Station. The city also sought to capitalize on its many educational institutions – College of the Holy Cross, Worcester Polytechnic Institute, Clark University, and Assumption University, among others – and in particular, to leverage medical services and education, building on the successful campaign to open the University of Massachusetts Medical School in Worcester in 1970, and the presence of the Worcester campus of the Massachusetts College of Pharmacy and Health Science University, opened in 2000. A cluster of downtown land parcels were cleared in the mid-1990s to build a new St Vincent's Hospital, which moved in 2000 from a predominantly residential neighborhood, where it had been located for more than 100 years. Its new location was a prominent mall-like structure built over a major north–south rail line, just a half mile from Union Station.

In the early 2000s, and throughout the next two decades (in spite of the economic downturn of 2008–2009), Worcester continued to redevelop its downtown in particular. The failed mall was torn down to make way for a new, more mixed-use development called CitySquare – parts of which cannibalized other sections of downtown by relocating existing employers to the new complex. For example, a major longtime (over 100 years) employer, insurance company Unum (originally named Paul Revere Life Insurance), relocated its workforce from a large, sprawling (and admittedly outdated) set of buildings on the western edge of downtown to a new building at CitySquare. Only seven years later, in the midst of the global pandemic of 2020, Unum announced that most of its Worcester workforce would not return to the office but continue remote work instead (this announcement led to some calls in the city for the tax incentives to Unum to be reduced, although no action was taken at the time) (Kotsopoulos 2020).

The CitySquare development in particular offers one form of evidence of a city successfully revitalizing. The development – a partnership between the city government and a real-estate investment group – includes office space, a hotel, and upscale apartments marketed especially to college students and young, upwardly mobile professionals as downtown residents and consumers. It was not the only downtown project in Worcester; Union Station, the iconic train station, became a multi-modal center as a bus transfer station was built adjacent to it, with all city buses stopping at that site. A new county courthouse was completed at the north end of downtown in the mid-2000s. A renovated theater, originally built in 1904, reopened as the Hanover Theater for the Performing Arts in 2008 at the south end of downtown. In one news story about the plans for CitySquare, a city development official cautioned that the project was not meant to solve all of Worcester's development challenges, but that multiple projects could together create a more vibrant, valuable economic center for the city (Palmer 2004).

All of these moves and investments reflect a common American municipal focus on expanding jobs, infrastructure, and ultimately the tax base in a downtown area. Most American cities – and many worldwide – fund critical infrastructure such as schools, roads (or road repair), water and sewer systems, police, fire, and parks departments through the local property tax base. These are usually supplemented by regional (state or provincial) government funding, but most US cities in particular rely on local property taxes for essential services like police, fire, and parks, and often for a large portion of their school budgets (see box, "Municipal Development and Finance Strategies").

Funding for CitySquare and other developments came in part from a common municipal finance tool: tax-increment financing (also known in Massachusetts as "district improvement financing" or "DIF"). Tax-increment financing uses the end goal of development, an anticipated increase in tax revenue from a new use, to fund the redevelopment itself. (In that sense, it is a bit like a leveraged buyout of a company, where the buyers use the anticipated future profits of the company they are buying to fund loans for the purchase price.) Typically, cities project an increased value of a completed development, and assess the "tax increment" that would be gained over a fixed period (usually 20 years). The city then issues bonds to pay for the development; the "tax increment" that is supposed to come to the city once the building is completed is diverted to pay the bonds over the 20-year period. Only after the tax-increment deal is complete does the higher tax value of the project accrue directly to the city to fund its usual obligations. Tax-increment financing is therefore "deferred value" from a municipal budget perspective, but cities pursue the strategy nonetheless in hopes of spurring additional private investment in downtown land uses and other tax revenue increases.

Many urban scholars have pointed to the ever-escalating competition among cities as they compete for development projects in hopes of raising their tax base value, adding jobs, and creating private investment. Private investors recognize that cities will fund property acquisition, leverage their own tax bases, and sometimes outright relieve tax burdens on private developers in exchange for property investment and job provision. So, cities that are seeking development (meaning, cities facing decline of their longtime industries

Municipal Development and Finance Strategies

City governments usually provide a range of services to their residents, sometimes in cooperation with private companies or community organizations (Smith et al. 2016). Typical municipal services include public schools, libraries, water and sewer services, garbage collection, police and fire departments, although the range of services provided depends on national context and whether other levels of government provide some services. In some places, residents do not get water, sewer, and/or garbage collection from their city government but instead have to pay private contractors for these services (Bakker 2010; Da Cruz and Marques 2011; Kirby, Knox, and Pinch 2017). In such instances, city governments often stopped providing those services because they could not do so and still pay all their bills, so they either subcontract to private companies (in which case people still pay the city directly) or they simply stop providing the service and instead expect private companies or community groups to step in. Where city governments stop providing these basic sorts of services like water and sewer or garbage collection, they may pass laws requiring that residents properly dispose of all forms of household waste, for example, as a way to force people to contract privately for these services.

Most city governments derive pay for their activities from two sources: taxes and fees. Municipal taxes usually come from property tax and also local sales taxes. Not all cities use both of these types of tax, but property value-oriented taxes are common enough that they connect to most urban development strategies: that is, city governments try to find ways to increase property development as a way to increase land values, and then to get a share of that increased value through taxation (Tapp and Kay 2019). Indeed, a significant urban development strategy in some countries is tax financing, where cities grant property tax breaks to companies that agree to locate within a city's boundaries, or leverage anticipated land

and property tax base) compete to give more breaks to developers or to lure potential employers, and these developers know that they can get tax breaks and other city assistance when they make investment decisions.

New downtown development does create "buzz" in the form of media attention. In the wake of the many developments in the 2000s and 2010s, Worcester was touted as a city rebranding itself. In 2015, for example, the *New York Times* celebrated Worcester's college town identity, acclaiming the landscape changes by saying "Long a College Town, Worcester now looks the part" (Schneider 2015). The article praises the city's marketing of the downtown landscape to the city's many college students (more than 30,000 across eight campuses):

> Worcester is attending to the 35,000 college students who study and live here, and its primary boulevards are steadily filling up with the civic amenities that attract new residents. They include a busy public transit hub, comfortable and affordable housing, new restaurants and watering holes, computer stores and coffee shops, a performing arts theater, biotech research facilities, incubators and office space for start-up companies, and renovated parks – including one alongside City Hall with an ice rink larger than the one in Rockefeller Center. (Schneider 2015)

The *Boston Globe* also promoted "the comeback of Worcester's downtown" in 2013 in a column that linked downtown growth to the innovation and youth of college students and the higher-education institutions that draw them to Worcester (McMorrow 2013). The themes of youth, activity, and creativity as economic drivers not only pervade media coverage of Worcester's downtown development efforts but also underlie many of the development choices themselves, with an emphasis on luxury housing for individuals or couples, restaurants, and entertainment.

Just as the city's drive for big development projects did not start with the CitySquare project, it didn't end there either.

value increases to pay for costs of redevelopment. The latter tactic is known as tax-increment financing or, sometimes, as district improvement financing. In both cases, a city will cover some or all of the costs associated with developing a low-land-value area (an abandoned factory, for example) by using the future increased taxes from a higher-value development to pay off the bonds or loans that were used to finance the redevelopment. So a city is borrowing future property tax revenue to pay for the improvements to an area that will create that property tax revenue. Conceptually, tax-increment financing is a bit like taking a loan to pay for college with the idea that, after college, a person will be able to get a job with a high enough salary to pay off the loan. But if a city bets on the wrong development (anticipating enough higher land value to pay off the loans with the increased property taxes), it might not be able to pay off its loans without raising other taxes and fees. Additionally, taxes may be "deferred" to pay for development for as much as 20 years, marking a considerable time frame – and building depreciation – before its higher taxes benefit urban residents in the form of more or better services.

Another dimension of the tax system is in how property owners, real-estate professionals, and investors respond to taxes and tax incentives (Weber and O'Neill-Kohl 2013; Tapp and Kay 2019). Land in a declining area means reduced taxes because the land is decreasing in value; then, over time, the property owner can realize tax incentives to participate in the redevelopment of that same land. Thus, landowners in cities have the perverse incentive to invest in devalued areas or disinvest in property so that it declines in value; eventually tax credits or tax financing will be available to subsidize the costs of redevelopment. These subsidies are usually forthcoming because city officials tend to operate on the assumption that they cannot expect private developers to pay for the costs of development without some form of tax break (despite the fact that it may be the overall tax structure and the presence of those tax breaks that delay investments in the first place, as owners wait for tax

Some city residents and developers had long targeted an area southeast of downtown known as the "canal district" (inspired in part by Providence, Rhode Island's riverside development about an hour away) for renewal, although the canal itself was long ago cemented over in Worcester. When the new owners of the popular Boston baseball team the Red Sox's affiliate minor league team in Pawtucket, RI (just outside of Providence) sought a new baseball stadium in 2015, some people in Worcester got involved in promoting the city – and specifically, an abandoned factory site at the north end of the canal district – as an alternative destination. For a while, it appeared that Worcester's gambit was destined to fail, as the "PawSox" owners sought a deal for a new stadium in Providence. But when state officials in Rhode Island declined to approve the public funding requested by the team owners to build either in Providence or Pawtucket, Worcester's interest – bolstered by a postcard campaign that enlisted Worcester residents in pushing for the move – shifted the focus to Massachusetts.

The postcard campaign offers an interesting interplay between the "power brokers" of city elites and everyday residents (Ballou 2018). The head of the business association in the canal district where the proposed ballpark would be located created a blank postcard promoting Worcester as the future site of the ballclub, and showed up with blank postcards for residents to fill out at street fairs, public parks, sidewalks, and so on. Eventually, more than 10,000 postcards were sent to the home of the Pawtucket Red Sox to promote Worcester. City residents who participated in the campaign by writing postcards likely imagined the promotion to be about luring a private business to the city – one with a link to the ever-popular regional major league baseball team, the Red Sox. It is not as likely that they actively concerned themselves with whether city or state tax revenues would be used to support this private enterprise.

In 2018, Massachusetts state officials, Worcester city leaders and developer partners, and Pawtucket Red Sox owners struck a deal for a new minor league ballpark to be built in Worcester between downtown and the canal incentives). In the United States, the assumption that cities have to entice developers with tax breaks is exacerbated by competition between cities for new businesses and developments (Molotch 1976; Harvey 1989a). Cities vie with one another to lower taxes to entice employers to locate there. It is also usually cheaper to build in an undeveloped suburban area – despite the higher environmental costs, such as loss of animal habitat and increased sprawl – than it is to redevelop existing built-up areas, so cities perceive that they have to subsidize costs. Sometimes these built-up areas have older buildings on them which are expensive to tear down (and even more expensive to rehabilitate) and which might have pollution in the soil or water systems. (Some cities do offer tax breaks specifically to entice developers to rehabilitate older buildings, however [Ryberg-Webster and Kinahan 2017].)

Cities may also impose local sales taxes if their regional and national governments allow them to do so. Sometimes when cities impose a sales tax, it can spur consumers to go to other towns, and especially suburban areas, to shop because they will not have to pay the additional small percentage of sales tax. Hotel and rental car taxes are sometimes significantly large (10–20 percent of a total bill) in large tourist-destination cities because tourists are not eligible to vote in those city elections and thus will not punish the local politicians for the high taxes. Local sales taxes affect everyone equally, but that means that poorer people, who have less income overall, will have even less, whereas wealthier people are not harmed by the relatively small tax on their purchases. Some scholars prefer property taxes because they primarily affect landowners but they may still impact poorer people in their rent costs.

See also related boxes "Urban Economic Processes" in chapter 2; "The Growth Coalition" in chapter 3; and "City Branding" in chapter 6.

district that would cost, all told, more than US$100 million. The ballpark itself would have a capacity for about 10,000 patrons. This time, the team owners achieved the public finance support they had failed to achieve in Rhode Island: the development, like others before it, was financed by district improvement financing, where the new buildings, including a hotel, apartments, and retail, are projected to produce enough new tax revenue to pay for the construction itself. The pandemic-induced economic downturn of 2020 undermined the rosy development assumptions that predicted that city taxpayers would not be on the hook for the costs, but construction on the stadium itself continued, and the ballpark opened in 2021.

Thanks to the payment of naming rights by a major local business, the new ballpark has been dubbed Polar Park; while its construction on a large empty site does not directly displace people, the development as a whole renewed concerns that development in Worcester focused on one type of resident – middle class, white, young – rather than a broader constituency of city residents. Indeed, some commentators in Worcester expressed concern about gentrification and displacement of existing residents from the two neighborhoods abutting the new development, where a number of the city's iconic multi-family three-story residential buildings are located, along with small-scale manufacturing and retail (Shaner 2018).

The narrative of downtown luxury, activity, and economic growth has a counter-narrative in Worcester, and highlights how the place of Worcester is itself multidimensional and exists at several different scales; that is, its identity connects people and processes at the hyper-local of a neighborhood or district (whether that be downtown or elsewhere in the city), and at the level of the state, nation, and global. These are the *relational place elements* of Worcester.

The counter-narrative is perhaps best illustrated by a 2016 feature article about Worcester in the *New York Times Magazine*, called "What Happened to Worcester?" (Davidson 2016). The article poses Worcester as the type of city that 100 years ago enabled new immigrants to find reliable jobs in factories. Yet the shift in the American economy from labor-based industry to a knowledge-oriented economy with mechanized industry – where the entry-level jobs are in services such as restaurants and retail instead of factories – undermines this accessibility to reliable jobs. While city leaders and regional property developers worked to transform the downtown and its adjacent abandoned factories into high end residential, office, entertainment, and retail-oriented uses, the rest of Worcester continues to labor in a wide variety of jobs, some of which offer unsteady hours and persistent low wages.

In the popular narratives of American newspapers, cities like Worcester are defined primarily by their job bases – factories, colleges, hospitals – and the people who fill the jobs in them. The key actors in downtown development are the developers and city leaders, both elected and appointed, who shape visions for "revitalization" and work with real-estate agents, developers, and financiers (bankers and city and state officials allocating development grants) to push for construction and renovation of downtown buildings. This development tends to emphasize consumptive-oriented economy: entertainment and services. Other actors who help drive downtown development, however indirectly, are the professional, educated, and often young residents or potential residents who fill the imaginary of developers and city officials as people who seek housing that is convenient to downtown transit and entertainment, and who will pay bigger rents for conveniences such as fitness rooms, security systems, and free shuttles to the train station.

Another imagined downtown "consumer" is the existing resident of the city's neighborhoods and the surrounding region, who would visit downtown if only there were new buildings, better restaurants, themed shops, and, especially, civic pride galvanizing entertainment such as a Triple-A Major League-affiliate baseball team. In a way, the desire to compete with suburbs still drives city officials to seek sources of consumer spending to "draw" people downtown.

The postcard campaign that helped to draw the Triple-A baseball club to Worcester offers a paradox and reminder of the always multiple and simultaneous experiences of and in a place. The effort by business representatives and some city officials to bring a Triple-A affiliate baseball team to Worcester clearly appealed to the many residents (more than 10,000 of them) who participated in the postcard campaign, perhaps because it activated a local pride and sense of possibility and vitality in the city. But the economic activity that the ballpark promised is largely entertainment, generally at lower levels of pay, and certainly not full-time year-round positions.

The key actors engaged in downtown-focused development are typically city officials, real-estate speculators and developers, banks, and local business leaders. Sometimes, residents actively oppose such deals because they fear the changes to the landscape and area culture that will occur, including rising rents if the development is successful. But city residents can also get excited by new prospects and growth, as evidenced by the postcard campaign for the baseball team. While people might care about how their downtown looks and feels, they might not feel they have a voice in the direction of the development. Or they might feel that writing a postcard to entice a nearby baseball team is the most that they can do. They just might have other issues on their minds: paying rent or a mortgage, ensuring the car is running, or dealing with the ice on the sidewalk from the latest snowstorm. There are so many multiple layers of life intersecting and competing in any given city on any given day, and we rarely explicitly reflect on how they connect.

Indeed, part of understanding downtown Worcester and its changes – and the chasing of the next development – is to recognize the ways that it is like other cities. Downtown malls, later razed to make way for mixed-use developments that include housing, are a story of 1950s–2000 American cities. Developers of downtown malls, and the subsequent new developments that followed, have a national perspective and usually a regional, if not national, business scope. Geographers have noted how urban planning and policy ideas are "mobile" as people transmit them around the world through policy papers, conferences, and practices (see box,

"Policy Mobilities"). Worcester too is connected to, and learns from, people in other places.

The city as environmental policy nexus

Like other cities, Worcester has other parts and experiences far beyond the conventional preoccupation with property development and downtown transformations. The framing in this chapter has focused so far on Worcester's most physically concentrated built landscape: the downtown, and the municipal economy of property development. This narrative assumes human activity but tends to concentrate on elites associated with elected and appointed government roles, and deep-pocketed investors, leaving out the concerns or patterns of everyday human life. Pulling back the lens to see more of the city brings a far broader range of elements and experiences into the urban landscape focus, including the non-human environment, such as animals, insects, air, water, trees, and grass; and the broader ecosystems of which they are a part (see box, "Urban Environments"). In Worcester, the broader environmental ecosystem has, at key crisis times, demanded human attention and focused on the *city as a place for nature*. One such time started in the summer of 2008, and while government actors at the state and federal levels played key roles, everyday residents affected by the environmental crisis also stepped into the action, asserting the importance of daily life in the experience of the city and questioning accepted wisdom about the best policy responses.

Worcester's natural environment became a site of specific governmental action and attention in 2008, when a resident found an unusual insect in a neighborhood near a large factory that once was the backbone of the local industrial economy and which now serves as part of a global conglomerate making industrial abrasives, using only a fraction of its tremendous physical plant. The bug was long and black, with white spots all over its body and distinctively long antennae. Known scientifically as *Anoplophora glabripennis*, it is also called the "starry sky beetle" where it originates in China, and the longhorned beetle (LB) (and sometimes "Asian longhorned beetle") in the United States. Its discovery in Worcester marked the start of a reawakening around the city's urban forest. Indeed, the discovery of the LB in Worcester marshaled an immediate and significant response that brought a team of foresters and government officials from a branch of the United States Department of Agriculture (USDA) called APHIS, the Animal and Plant Health Inspection Service. Because the spotted beetle with the unusually long antennae was not native to the United States, it had no natural predators. It burrows and plants its eggs in, and feeds off, hardwood trees, eventually weakening and killing them.

The energy and attention to trees that the finding of the LB engendered was only intensified when an ice storm affected the city in mid-December of that year, and countless tree limbs, possibly weakened from the feasting of the LB, came crashing down, especially in the higher-elevation northern neighborhoods in the area where the beetle had been found. Schools were closed for almost three weeks while the streets and sidewalks were cleared of tree limbs, and power was slowly restored to the region.

These events, the discovery of the invasive beetle and then a dramatic weather event highlighting weakened and vulnerable trees, marked the beginnings of a campaign, waged at first by the federal and state governments and then also by city residents, to cull and then plant new trees throughout the city. The initial focus was on eradicating the beetle and stopping its spread in the region; state and federal officials argued that if the beetle were allowed to live unfettered, the New England forests as a whole could be threatened, since LB lived in and feasted on hardwood trees, especially maple trees. In Worcester, the predominant species of maple where the LB was found was itself a non-native "invasive," the Norway maple, planted in large numbers after a tornado affected the city in 1953. The same neighborhoods of the city that had been devastated by the tornado were affected by the LB eradication effort, again creating a landscape denuded of trees.

> **Policy Mobilities**
>
> Mobility generally invokes an idea of movement, and thus perhaps fosters imaginaries of people (or maybe animals) in motion. But scholars also emphasize that ideas and policies move. Specifically, in urban research, policy mobilities refer to the way that ideas about urban economies, planning, development, and redevelopment travel from one place to another, standardizing to some degree the urban policy "toolkit" worldwide (McCann 2008; McCann and Ward 2011). Policy ideas circulate through the interactions of people, such as at planning conferences or meetings of mayors and other elected and appointed officials, or via "best practices" of practitioners like international organizations (McFarlane 2009; McCann 2011; McNeill 2011). Yet, rather than wholesale transfer of policy from one place to another, it might be more accurate to think of policy mobilities somewhat more flexibly, as urban planning and development principles are adapted across different urban and national contexts (Peck 2011). Mobile policies include strategies like bidding to host the international Olympics, which cities then use as a means to develop significant "event" infrastructure, such as arenas and sports facilities, and to spur international tourism (Lauermann 2016; Müller 2017).
>
> See also related boxes "The Growth Coalition" in chapter 3 and "City Branding" in chapter 6.

The discovery of the LB in Worcester was not the first in North America. The beetle had also been found in cities such as New York City (Brooklyn, in 1996), near Toronto, Canada (2003), and Chicago, IL (2008) – in all cases, traced to global shipping, with wooden pallets being the likely source of beetle larvae that then emerged in new locales. The strategy taken by APHIS and the USDA more broadly in every case was to cut down any trees found infected or in proximity to infected trees; this same approach was undertaken in

Worcester, where APHIS officials worked with Massachusetts state authorities from the Department of Conservation and Recreation. Between 2009 and 2012, over 30,000 trees were removed from neighborhoods in Worcester and surrounding towns in the LB quarantine zone, felling trees in both private yards and along streets in the northern Worcester neighborhoods of Burncoat and Greendale.

Worcester residents, especially in the most affected neighborhoods, roiled in protest at the destruction of their treescapes. Some urged city, state, and federal officials to find insecticides that could control the beetle without cutting down trees. Officials argued that such efforts would not adequately control the beetle and insisted that, if left unchecked or insufficiently controlled, the LB could spread across all of the New England forests and into Canada, destroying not only the trees but the tourism and maple sugar economies with them. Official APHIS and state policy remained unchanged: stop the LB in Worcester by destroying its local habitat. The identity and character of the local neighborhoods in Worcester were subsumed by concerns expressed at much broader levels; the infested urban forest of Worcester was seen as a threat to the New England forests of an entire region. This broader scale shaped the local response, in part because it was federal and state authorities that dictated (and funded) it. But it was also the broader concerns and the involvement of federal officials in seeking to eradicate the beetle infestation that enabled a policy response that refocused attention on the lost treescapes.

The policy that emerged addressed the local concerns about tree losses while still prioritizing eradication of the beetle, mostly through tree cutting. One of the key policy actors at the federal level was Congressman Jim McGovern, who had represented Massachusetts' Second District in Congress since 1996, and who grew up and continued to live in the Burncoat neighborhood of Worcester – one of the hardest hit by the LB infestation. As a resident of the affected neighborhood and also a member of US Congress, he could easily understand both of these two narratives about Worcester in the wake of LB: a regional and national concern about forests; and a

> **Urban Environments**
>
> Cities and urban places tend to foster imaginaries of people, buildings, streets, and human activity. There isn't a lot of room in the typical urban imaginary for trees, plants, water, and truly fresh air. And yet a significant amount of urban thought has been, and continues to be, dedicated to considerations of the natural environment. As early as 1870 in the United States, Frederick Law Olmsted wrote about urban parks as healthy sources of fresh air for city denizens (Olmsted 1870). His focus and framework was the industrial city, teeming with pollution and overcrowded housing, and didn't really consider nature beyond the tamed and organized spaces of formal parks – which were highly manicured.
>
> In his book *Nature's Metropolis*, historian William Cronon unpacked and problematized the presumed distinctions between urban and rural, built and natural environments. He argued that "country" landscapes were actually part of urban places because rural economies were fully imbricated through the extractions and transformations of land, water, and forests into economic uses in cities. This framing acknowledges the different types of landscapes in cities as compared to towns, farms, or forests, but encourages thinking about urban places as also incorporating the biophysical environment. Matthew Gandy (2002) extends this thinking, arguing that the development of New York demonstrates the emergence of a "metropolitan nature" which characterizes the management, transformation, and use of nature (such as water, parks, forests, extracted resources, etc.) in urbanization generally (see also, for example, Huber and Currie 2007). Other scholars explore similar themes about urbanized nature, whether it is physically located within cities (Anand 2017) or as far away as Antarctica (Brenner and Schmid 2015); the point is that nature fully supports and enables urban living and, also, that urban living irrevocably changes and affects natural environments everywhere.

local outcry about landscapes altered by massive tree cutting. He participated in an effort by local officials, policy makers, and other Worcester residents to create a local organization to encourage tree planting. Officially founded in January 2009, the Worcester Tree Initiative (WTI) was created for one purpose: to replace the 30,000 trees lost in Worcester to the anti-LB campaign over the next five years, from 2009 to 2014.

The Worcester Tree Initiative fostered tree planting through tree giveaways, where residents or other property owners could come to events where juvenile trees were given away in exchange for pledges to plant and care for the trees. Any property owner within the APHIS-designated LB quarantine zone – which included all of the City of Worcester, and significant parts of surrounding suburbs – could get a free tree. At the same time, the City of Worcester allocated resources to planting street trees, particularly in neighborhoods where trees were cut down or that had never had significant tree canopy. The State Department of Conservation and Recreation (DCR) also initiated its own tree-planting program, targeting low-canopy areas in Worcester and employing foresters to plant the trees. So while federal intervention induced a tremendous tree-cutting campaign to stop the LB from spreading beyond Worcester, it was also federal investments that enabled the city and the state of Massachusetts to refocus on trees at the local level, remaking city landscapes as they did so.

Worcester as an urban forest

The narrative of Worcester that emerges from the discovery of LB in 2008 is one of renewed attention to at least one aspect of an urban physical environment: its trees and the urban canopy formed by them. When asked by researchers about the impacts of the LB infestation,[1] residents who lived in the most affected areas where trees were cut down expressed appreciation for trees now lost, or lamented the changed landscape:

Another line of research on urban ecologies examines biophysical processes in cities focusing, variously, on pollution (Nowak, Crane, and Stevens 2006; He, Pan, and Yan 2012), soil and water quality and systems (Bakker 2012; Swyngedouw, Kaika, and Castro 2002; Pouyat et al. 2015; Palta, Grimm, and Groffman 2017), trees and parks (Ibes 2016; Nath, Han, and Lechner 2018), and the importance of cities to be environmentally sustainable in light of global climate change (Cohen 2017; Verma and Raghubanshi 2018). In the area of urban forests, for example, scholars examine how the tree canopy helps to mitigate the effects of higher temperatures caused by the impervious surfaces of densely built urban environments (de Vries et al. 2003; Donovan and Butry 2009; Elmes et al. 2020). They also examine the unevenness of urban forests, where some urban areas have far more trees than others. These are usually residential areas with single-family homes, where trees are especially compatible with the landscapes (Heynen, Perkins, and Roy 2006; Locke et al. 2021). Research on uneven tree canopy highlights an important line of investigation for some urban environmental scholarship, namely, environmental racism (e.g., Bullard 1993; Pulido 2000, 2016). Environmental racism describes how social and political systems, including structural racism, produce landscapes in which people of color face far higher exposures to pollution and unsafe living conditions, and far less to healthy green space.

See also related boxes "Ecomodernism" in chapter 5; and "Landscapes and Power" in chapter 7.

[Trees] have so many benefits, right, not only just their beauty but clean air, they keep our houses cool, keep our houses safe from bad windstorms.

I mean they're holding the soil together, they're creating oxygen. They're shading our houses. They're giving us peace.

Maturing trees shape a new streetscape. The tallest tree, front left, is the only one abutting the street that was unaffected by the longhorned beetle. The rest of the trees along the street are only about ten years old, having been planted after all the other mature trees were cut down to prevent the spread of the beetle[2]

Now it looks like a bowling alley. And people get, people are crying, people get emotional about that stuff.

You drive down the street where there once were huge trees, then nothing.

The Worcester response to LB eradication also fostered significant environmental policy changes. The replanting programs – which successfully met their 30,000 tree-replacement planting goal in October 2014 – launched a new era of tree care, stewardship, and attention in Central Massachusetts. Although it was founded initially as a five-year effort focused just on replanting the number of trees lost to LB eradication, the WTI is now established as a component of the outreach programs of a regional organization, the Tower Hill Botanical Garden. The organization continues to encourage tree awareness and stewardship through its education and planting programs. While residents in the northern Worcester neighborhoods of Burncoat and Greendale were hard hit by LB and the winter 2008 ice storm, resulting in significant tree loss and neighborhood change, residents in lower-lying parts of Worcester might not have even known that LB and its federal response was happening.

Lower-income and more ethnically diverse southern Worcester neighborhoods had less tree canopy before the LB infestation even occurred and were largely unaffected by tree cutting. When the state joined in the tree-planting effort, it focused some attention on Worcester city streets where the tree canopy had long been neglected. In doing so, it created a new tree-planting program which involved state foresters in planting trees, working with local residents and property owners to care for the trees (especially watering in the first few years). This pilot program eventually became a statewide initiative called the Greening the Gateway Cities program, where a state agency plants trees in "environmental justice" neighborhoods of small and mid-sized Massachusetts cities that have less access to green space and urban tree canopy benefits like summer shading and winter windbreaks.

Worcester as a place that experienced and responded to the presence of LB works at multiple levels of individual and community action and geographical scope. It is a city where the beetle could appear because of its imbrication in a global industrial economy with goods shipped on wooden pallets from halfway around the world. A city where someone noticed an unusual bug. A city where residents responded with anger and suspicion to government agencies that sought to destroy a beetle by removing thousands of trees. A city where powerful individuals and agencies, including a federal Congressman, established new organizations that connected with people at the grassroots, in tree-giveaway events and through increased tree planting in low-income neighborhoods.

The LB infestation and response oriented people to the physical environment but in different, albeit overlapping, ways. Politicians and federal agencies focused on insect eradication, while residents reflected on the ways that trees shape their landscapes and even their social lives. For residents, the loss of trees registered first and foremost at the intimate scale of home and neighborhood, and the focus was primarily on trees in both emotive as well as financial (energy costs) registers. At the same time, the infestation response of the federal government – to cut down trees in order to stop the spread of the beetle – highlighted for residents and policy makers the important location of Worcester in Central Massachusetts in relation to the broader New England forests, which comprise a key component of the regional economy. The LB infestation and response highlight that Worcester is more than just its downtown or economic parts; it blends a human-built environment and a physical environment through actions and relationships which extend and interconnect far beyond the city's official boundaries.

The successes of the tree planting – out of a frustrating and concerning situation – highlight one aspect of Worcester's place identity, that of a creative, networked response to adversity. While all of Worcester was within the LB infestation zone designated by the USDA, not all of Worcester was equally impacted or involved in the response. The attention by the state – which planted trees in all six municipalities in the LB infestation zone – to areas untouched directly by LB again highlights the ways that Worcester events and actors reverberate on a broader scale, connecting Worcester residents and events to statewide, or global, events and initiatives. So while the vast number of actors in the WTI planting response was in Worcester's more advantaged northern and western neighborhoods, the planting in low-canopy areas enrolled a greater percentage of residents, indirectly, into the planting response.

Conclusion: the place/s of Worcester

The longhorned beetle infestation and subsequent tree-planting efforts represent a very different place frame from the downtown frame which highlights economic growth and development. Both involved significant local, state, and even federal actors and resources but they affected residents, workers, and visitors quite differently, if at all directly. Both are equally "true" yet always incomplete representations of the city. Some actors – notably city, state, and federal officials – overlap and connect the stories. In doing so, they bring multiple experiences of Worcester, its economy and built environment, residential areas, and physical landscapes, into multiple policy narratives. In these framings of the place, some elements are highlighted while others are downplayed or ignored, while policy makers at different levels of government stand out as key actors. "Local" events and people, then, also always extend to other sites and realms of action, sometimes in ways that obscure the immediacy and experiences of the local.

Worcester's economic policies, as seen through downtown development, emphasize efforts to boost the tax base and attract consumption-oriented businesses such as high-end residential, retail, and entertainment uses. These initiatives have similarities with both discourse and policy in other places. They also have real effects for area residents, such as rising rents or changed access to services. One of the major physical reorganizations of downtown, for example, moved the bus-transfer site from City Hall – where buses clogged Main Street and customers had easy access to shops – a few blocks away to the train station, creating a multi-modal, dedicated bus station. But the walk to stores and City Hall got longer. Focusing on downtown and tracing the links, changes, and motivations of any single development or initiative requires attention to multiple actors and impacts beyond those actors. But any single frame, no matter the depth, will then miss other elements, such as the trees along the streets and sidewalks, mostly ignored until a part of the city faced losing a significant number of them. Shifting the lens, then, to the trees, the city's urban canopy, and its green neighborhoods brings a different set of concerns, actors, and motives into view.

These representations of Worcester capture parts of the whole: the city as an amalgam of built infrastructure, natural environment, and the many people who live, work, or just drive through the place on the way to live, work, or visit someplace else. Understanding cities as this mix requires thinking about major and minor narratives of what makes up a city: density of people, ideas, and activity; economic exchanges; human–environment intersections; conflicts; and interactions. We think about Worcester, and we think about cities, when we consider these various elements and how they intersect, conflict, coexist, and extend beyond the place itself and to other places. Worcester reminds us of a hyper-focus on economy, tax base, and growth, but also that these happen while ecological processes are also ongoing and transforming the landscape. Sometimes those ecological forces demand attention, usually in a crisis. These two themes demonstrate the disparate yet interconnected time–space trajectories of cities-as-places, and point to the many other place frames to consider and unpack in order to know this place even more.

This chapter might help you to think about cities like Flint, MI, United States; Las Vegas, NV, United States; Miami, FL, United States; Sheffield, United Kingdom; and Novosibirsk, Russia.

–5–
Portland: Paradise of Environmentalism/Legacy of Exclusionary Racism

Portland, Oregon, United States, is a city at the center of an urban region of 2.5 million residents. Over the last 30 years, it has been identified as a leader in the United States in so-called "green" urban development strategies. Local and regional governments have poured money into public transportation, invested more broadly in green infrastructure, and (as a result of pioneering legislation passed in 1973) created a regional urban growth boundary that discourages sprawl by prohibiting suburban-style development at the urban fringe.

Many residents of Portland are quite proud of the city and its region. They often characterize it as a model of sustainable urban development in the US context. Some hail the past several decades of slow but steady densification and consistent economic growth as offering important lessons to other cities around the nation and (to a far lesser degree) the globe.

The city's culture is distinctly left-of-center in the US context. It is self-consciously quirky, making it a subject of both good-natured parody from the left (as in the television show *Portlandia*) and sharp-elbowed critique from the political right. For many on the political left – especially the environmental left – Portland (while obviously imperfect) stands as a reference point for how to put urban political words into action through a sustained, reflexive commitment by city and regional governments to policies motivated by ecological ideals over a period of decades.

Narrating Portland in this way requires a framing of Portland as a city of the United States rather than of the globe. There are many cities around the world that are significantly denser, more transit-centric, and more committed to policies promoting ecological sustainability than is Portland. Furthermore, this view of the city significantly avoids the existence of Vancouver, WA, just across the Columbia River. Over the past several decades, Vancouver has acted as a kind of escape valve for a number of browner, less dense development practices that have contributed materially to the combined metropolitan area's economic growth.

Perhaps more uncomfortably, the widely held history of Portland as a kind of green mecca for the American urban left also sits uncomfortably against the city's (and the state of Oregon's) long parallel histories of explicit white supremacy, racial exclusion, and colonial violence against native peoples. Framing Portland as a left-wing idyll means telling the Portland story in a way that begins no earlier than the 1950s, obscuring the lives of people who were excluded or annihilated before that time. It also means minimizing the role of present-day racism in the ongoing process of remaking the contemporary landscape of the city.

We start by tracing a conventional place frame of Portland as *an icon of ecomodernist progress*. After establishing this frame, we compare it to a frame in which *modern development is a continuation of a history of racial and colonial erasure*. For most of the last 50 years, the ecomodernist frame has dominated discourse in Oregon. However, the protests in Portland following the George Floyd killing in 2020, as well as the federal policing strategies in the city toward the end of Donald Trump's presidency, have contributed to an unsettling of this ecomodernist frame that to some degree recenters racial oppression in the city region.

Portland as an icon of ecomodernist progress

World War II jump-started a period of tremendous investment and growth in the Portland region. Between 1940 and 1945, three major dockyards in the region (including two within the municipal limits) produced over 1,000 "Liberty ships" in support of the war effort. Portland before the war was a tidy, modest city that was dwarfed by Seattle to the north and the even larger, more important economic center of San Francisco to the south. Oregon's identity in the early twentieth century was still as much "American frontier west" as it was "American west-coast culture." World War II significantly changed the city's trajectory.

As the population of the Portland metropolitan area continued to grow rapidly in the 1950s and 1960s, farmers and residents in the rural Willamette Valley to the south of the city looked on in dismay. The Willamette Valley is a productive farming region with mild temperatures, excellent soils, and easy water access. Due to the physical constraints of the region's geography, however, it was clear to stakeholders throughout the state that, without intervention, suburban sprawl from the growing city would be channeled south along the Willamette River and the agricultural lands surrounding it. Mountains hem in potential growth to the east and west. Were the status quo pattern of development to continue, within a few generations farming would no longer be a substantive industry in Western Oregon. The ruralist lifestyle and cultural patterns that residents treasured would be gone.

At the same time, the so-called "counterculture" of the 1960s was overflowing from San Francisco up the coast into places like Portland. The counterculture was a left-wing cultural and political reaction to postwar conservatism that developed most rapidly in major population centers on both coasts of the United States in response to the conservative, nuclear-family-centered national moralizing and anti-communist politics of the 1950s. These new left-aligned Portland residents believed in the social benefits of urbanism and looked down upon increasingly prevalent suburban development patterns that they perceived as backward or even de-humanizing.

In the early 1970s, an unlikely political coalition formed. On one side stood the so-called left-coast hippies and allies who favored dense urban environments in Oregon – primarily in Portland, the state's largest urban center. On the other, conservative ruralists and agricultural interests wanted to defend their agricultural activities. Both groups saw it in their separate interests to bottle up urban development at the northern end of the Willamette Valley. This coalition organized and lobbied toward the passage of a new state law, SB100. This law required that every urbanized area draw an urban growth boundary to spatially constrain future development. After several years of negotiation and the creation of "Metro," a supra-municipal layer of government to manage regional development activities, a newly defined growth boundary was enacted in the Portland metropolitan area in 1980.

The urban growth boundary (UGB) is not static. It has been expanded dozens of times since 1980. In fact, Metro is legally required to maintain room for 20 years of future planned growth inside of the boundary. This means that Metro has to plan not only for future development in general, but to imagine the increasing future density of development patterns across multiple local municipalities. The more it is the case that future development increases density, the less undeveloped land is needed inside of the boundary. So Metro's role is unusual in the United States: it decides, decades ahead of time and across multiple governmental units, where increases in density will be required in future urban development if that development is to continue.

Over time, the UGB has become core to the region's identity of responsible, green urbanism. As intended, the city-region's ongoing growth in population has been coupled to densification instead of new sprawl. Additionally, Metro and its constituent municipalities have invested heavily in a downtown-centered streetcar system, a regional light-rail system, and (to the west) a commuter rail line, with

Sprawl, Density, and Urban Growth Boundaries

After World War II, the rise of the automobile in the Global North came hand in hand with suburbanization (Jackson 1985). Automobility invited radically new patterns of urban development. Instead of fostering high densities near key modes of shared transit – or even walkable commutes in the urban core without engaging with transit at all – cars induce landscapes of moderate or low densities that disperse users in space and time. The shorthand that urban scholars (and practitioners) use for moderate-density development centered on automobility at the urban fringe is "sprawl."

While "the suburbs" are unironically appreciated by some residents, the term "sprawl" is usually associated with critique (Rusk 1993; Gillham 2002). Because cars are a very low-density form of transit, car-centric landscapes induce congestion even at passenger loads that are very low compared to buses, trains, or even walking. They require extensive space for parking anywhere that many people expect to go, even in the city center. As a result, car-centric landscapes are by necessity much more diffuse and perforated than those oriented toward other modes of mobility. The archetypal new public spaces of the mid-to-late twentieth century (for example, the shopping mall or the theme park) turn inward on themselves to facilitate a ring of parking around the outside, separating themselves from any kind of neighborhood fabric.

Urban growth boundaries (UGBs) are a political intervention to limit sprawl when economic or social interests otherwise will not (Easley 1992). There are different legal models – some use a belt of public parkland at the urban edge to discourage suburban use beyond, while others simply mandate particular kinds of use through zoning or similar legal requirements. Oregon is not the only US regional state with a growth boundary, and indeed urban regions in France passed UGB laws earlier, in the 1960s. Growth boundaries are a strategy to incentivize and facilitate coordinated transit-oriented design projects at major nodes throughout these networks. The city has simultaneously made promises not to expand highway capacity in the region, driving commuting toward public transport.

The light-rail system has been the backbone of a multi-decade commitment to densifying around mass transit. As the regional light-rail network has been built out since the late 1990s, it has been accompanied by corridors rezoned for higher densities and mixed use. Ridership has been strong for a city of Portland's size, and the regional government continues to plan for decades of significant network expansion.

The dramatic, lush Pacific Northwest forests surrounding the city, combined with the rejection of low-density suburban development patterns, the commitment to preserving regional farm capacity in an era of rising interest in local food, and the extensive visible investments in public transportation, make Portland a national icon of responsible left-wing development. In the 1990s and 2000s, left-leaning college graduates chose to move to Portland at unusually high rates, making the city both unusually young in its demographic profile and relatively highly educated compared to statistically similar urban areas.

Today Portland is, in its own terms, an ongoing success story of progressive environmental urbanism. The dominant place frame in Portland defines the city as a quirky ecomodern mecca. Echoing the slogan of Austin, TX, stickers posted across the city exhort residents to "keep Portland weird," where *weird* signifies characteristics like openness to experience, openness to collective and/or state action, and a willingness to make both individual and collective choices that run against the grain of America's great suburban expansion from the 1950s onward. Portland's explicit collective identity is nonconformist, environmentally oriented, pro-urban, and leftist.

While in some cases the urban environmentalism of residents is carefully articulated, more often it is an amorphous affiliation. Words like *environmental, green, sustainable,* and *eco-friendly* are often used to characterize the city by people who then struggle to identify anything especially green or

> achieving over time the levels of core density that support more efficient and more environmentally friendly landscapes in the future.
>
> See also related boxes "Policy Mobilities" in chapter 4; and "Land Markets" in chapter 6.

environmental about their daily lives (Pierce 2011). Green urbanism is often experienced more as a kind of naturalist engineering aesthetic: shaggy berms topped with long grasses in the parks, or engineered-timber structures with exposed structural joists and large expanses of UV-managing glass. Yet however unclear the concept of greenness might be, many Portland residents embrace the city's association with it (Pierce 2011).

The existence of the UGB means that the spatial footprint of Portland's metropolitan area is unusually clearly defined. It is situated at the northern end of the Willamette Valley where the Willamette River meets the larger Columbia River, just before the Columbia finishes its route to the Pacific. While adjacent municipalities within the growth boundary are clearly lower density than the city center, they are more sharply defined in contrast to the pastoralism of the lands just outside of the boundary than against the higher density of the core. The UGB clearly "works" in the sense that it both promotes density and preserves rural land uses in close proximity to the city.

Portland's compact downtown area is physically rendered in sharp relief by steep hills to the south and west and the river to the east. As a result, it is linked to surrounding neighborhoods by a set of bridges (across the river) and only a few major roads (to the southwest). Like Pittsburgh, PA, another city set along the confluence of rivers and surrounded by hills, Portland's downtown is physically isolated, connected by relatively few heavily used bridges, transit, and major arteries to neighborhoods with which it lacks contiguity.

The downtown cluster is genuinely dense by the standards of medium-sized cities, but the pro-density rhetoric for the

Public transport connecting downtown Portland and nearby Beaverton[1]

region as a whole shows a somewhat strained relationship with reality. The region's population density is roughly equivalent to that of notoriously car-centered Dallas, TX, or notoriously postindustrial (and depopulated) Detroit, MI. Nevertheless, the Portland region is clearly committed to densification even if it has not (yet) achieved a condition of systematic density.

There are a lot of performative displays of both greenness and "weirdness" in the urban landscape. Sometimes these are combined: a tall downtown tower with a rank of tiny wind turbines around the rooftop, or a Rube Goldberg gravity-fed irrigation system for a front-yard urban farm in a residential district. But just as often the weirdness and the greenness are distinct. Funky local standby restaurants have intricate New Age visuals but often don't really engage in visible practices of sustainability. Similarly, as a wave of new construction remade the center city from the 1990s on, a sleek green-glass ecomodern aesthetic ruled the day. However, the aesthetic often isn't especially compatible with the idea that Portland is "weird" or funky.

Ecomodernism is a label for an aesthetic and a philosophy that embrace contemporary technology and a progress narrative as part of "solving" the problem of environmental degradation, non-sustainable societies, or climate change. It is defined in contrast to a kind of "back to the earth" environmentalism that rejects modernity as the problem. As emblemized by the growth boundary and the light-rail buildout, Portland's civic culture is of a "do something" environmentalism rather than a "stop doing something" one. This is a city excited about the coming wave of engineered timber skyscrapers, not a city of people looking to leave cities and live a low-intensity life on rural farms.

The growth paradigm may be incrementally toward less destructive patterns, but it is a growth paradigm nonetheless. In many – most – ways, Portland has a very conventional development narrative that happens to be situated in a lush, verdant forest region. The urban area's most significant commitments to an ecomodern landscape (development through densification) were historically not primarily motivated by a sustainability ethos but instead by a coalition between people with agrarian aesthetics and those with urban ones.

Portland as a place of white supremacy, colonial land theft, and racial erasure

The ecologically sensitive place narrative traced above is overwhelmingly reproduced by white and self-identified progressive Portland residents. Portland is one of the whitest major cities in America (with over 75 percent of its population defined as white in 2020). But *why* is Portland such an unusually white city in the American context, especially compared to its port-city neighbors to the south (Los Angeles, San Francisco, and Oakland) and to the north (Seattle)? The answer can be found by grounding present-day development in a longer history of Portland and the regional state of Oregon surrounding it, one that attends to the explicit white supremacy of the region's early white colonists.

Beginning when white colonial settlers moved into the territory in significant numbers in the early to mid-1800s, the laws of the colonial state explicitly promoted the exclusion of Black individuals, as well as excluding native communities from the full protection of its laws. In the period before Oregon's admission to the United States as a regional state, white settler residents came together to form a provisional territorial government. As was true of other white settlers in other US territories, they hoped that this act of self-organization and ongoing self-regulation would galvanize the US federal government to begin the process of creating a new state. These white settlers, only numbering roughly 150 white men of American and French-Canadian backgrounds, voted to create such a provisional government, along with a set of "organic laws" (passed in 1843, revised in 1845) to serve as a basic legal framework for the inchoate territory.

In the summer of 1844, the provisional government passed a law that forbade slavery in the territory. However, the law also forbade Black people from residing within it. After an

Ecomodernism

The modern environmental movement's roots are often located in late nineteenth- and early twentieth-century admiration of wilderness landscapes. John Muir (1901) and Henry David Thoreau (1854) wrote aesthetic and philosophical treatises on the intrinsic value of nature and of human beings' encounter with it. During the 1960s and 1970s, awareness of environmental risks and activism about them rose significantly across the Global North. Rachel Carson's book *Silent Spring* (2002 [1962]) was a catalyzing text, describing the effects of synthetic pesticides on humanity. Key laws protecting natural lands and regulating environmental toxins were passed in northern countries in the middle part of the twentieth century.

During this period, environmentalism was often antimodern in character. The core of the environmental critique was that unchecked population growth and the technological expansion of ecological footprints were inherently damaging to the non-human environment, threatening the carrying capacity of Earth (Ehrlich 1968). This mode of environmentalism is in a certain sense conservative: it seeks to counter the tendency toward radical new technologies by emphasizing the historical logics of, and relationships with, the land. This conservationism is sometimes in conflict with political conservatism, which seeks to protect and extend existing patterns of social and political power; but both revere older practices and are skeptical of new patterns as disruptive of safety and order.

Ecomodernists, in contrast to the conservationist movement, argue that the way to save a healthy and functioning ecological system under pressure is through the expansion of new technologies (Brand 2009; Asafu-Adjaye et al. 2015). Rather than abandoning modern science, ecomodernist advocates seek to speed up the pace of development of new environment-saving machines for living. These investments include solar, wind, and "safe" nuclear energy; carbon-sequestration technologies to manage and initial grace period, the law required all Black and mixed-race individuals to leave the territory and never return; any Black individual remaining would be subject to up to 39 lashes from a whip, repeated every six months, until that individual left the territory. In December of 1844, the legislature thought better of the severity of the law's penalties and made revisions. Instead of lashing, any Black person remaining in Oregon would be subject to arrest and hard labor, followed by bodily expulsion.

In 1849, still fearing the possibility of an influx of new Black residents, the territory passed a law requiring any Black person entering the territory to leave within 40 days.

In 1857, as part of the run-up to statehood, Oregon ratified its new state constitution. Oregon entered the union as a free state (i.e., without the institution of slavery), but the constitution included the following text in its Section 35:

> No free negro or mulatto not residing in this state at the time of the adoption of this constitution, shall come, reside or be within this state or hold any real estate, or make any contracts, or maintain any suit therein; and the legislative assembly shall provide by penal laws for the removal by public officers of all such negroes and mulattoes, and for their effectual exclusion from the state, and for the punishment of persons who shall bring them into the state, or employ or harbor them.

Oregon was the only state to be admitted to the United States with a constitutional or other legal provision excluding Black people from its lands. Section 35 of its state constitution was superseded by the passage of the 14th Amendment to the federal constitution of the United States in 1868 (after the conclusion of the US Civil War) and so rendered moot. However, Section 35 was not formally repealed until 1925, and other obsolete but explicitly racist language (for example, setting thresholds for changes to state policy based on the number of *white* residents in the state) remained in the state constitution until removed through a statewide ballot initiative in 2002. In that referendum, 29 percent of Oregon

> reverse climate change; and less-polluting transport strategies like electric vehicles and high-efficiency mass transit. Ecomodernists believe that the problem isn't technology per se, and that it is a mistake to give up the benefits of modern science and technology when better technologies could provide the best of both worlds.
>
> Ecomodernism and traditional environmentalism both seek a cleaner, healthier ecological context for humans, but they seek it in very different ways and mobilize distinct political alliances to try to move toward the better futures they seek.
>
> See also related boxes "Modernism and Urban Design" in chapter 2; "Urban Environments" in chapter 4; and "Landscapes and Power" in chapter 7.

voters chose the option that would retain the obsolete and legally unenforceable, but explicitly racist, language (State of Oregon 2002).

Even though the Black exclusion laws were rendered obsolete in 1868, Oregon continued to be a fiercely inhospitable place for people who were not white. The Ku Klux Klan was a powerful cultural force in the 1910s and 1920s, when the Portland chapter boasted 16,000 members at its peak (Bruce 2019). As late as 1940, when the population of Portland had risen to over 300,000, it was still the case that only roughly two thousand Black people lived in the city. City properties like pools and playgrounds remained segregated by race through into the 1960s (Stephan 2017).

During World War II, a private company owned by American industrialist Henry Kaiser recruited Black Americans to work building liberty ships in Portland. Kaiser's company organized the building of new publicly funded housing in a brand new municipality, Vanport, on reclaimed land by the banks of the Columbia River and adjacent to the dockyards. Nearly all of the new Black residents of the Portland region were settled in Vanport from 1942 onward. The population of the historic city of Portland remained almost entirely white throughout the wartime. This was not accidental, but rather an effect of ongoing real-estate discrimination and explicit racial threats in the city of Portland proper.

Shortly after the war ended in May of 1948, a berm separating Vanport from the Columbia River failed, and the entire city was flooded within a matter of hours. The roughly 18,000 residents of Vanport, a majority of them Black people, were rendered instantly homeless.

Many of those Black residents once again left the inhospitable territory of Oregon. But many of those who stayed resettled a few miles away in Portland's first recognizably Black neighborhood, Albina. Albina was never an exclusively Black community in the way that many other American cities had ghettoized areas – there simply were not enough Black people in Portland to dominate even one neighborhood in this way. However, like those other overwhelmingly Black neighborhoods in other cities, Albina suffered from systematic disinvestment and a lack of municipal services from 1948 onward. It was repeatedly targeted for "urban renewal" and "slum clearance," the mid-twentieth-century technocratic parlance for remaking the urban landscape that most often masked the project of removing Black residents from urban terrain (see box, "Residential Segregation and Redlining in America"). Key strategies included the early removal of streetcars from the city's network and then the construction of a major highway directly through the neighborhood. Only in the 1990s, when the city began to build a new light-rail line parallel to the highway, did new investment, services, and (classically) displacement of Black residents from this historic center of Black life in the state ensue.

In the contemporary era, Portland remains one of the whitest large cities in America. As of 2010, Black residents constituted only 2 percent of the population of the state of Oregon as a whole, 6 percent of the city of Portland, and less than 50 percent of the King section of Albina, the district with the largest non-white population in the city.

This long history of racial exclusion has been largely absent from the dominant ecomodernist place frame of the Portland region. The hegemonic left-leaning self-narrative of the city is

Residential Segregation and Redlining in America

Cities manifest social differences visibly in the landscape. Although housing styles vary architecturally around the world – because they reflect the cultural influences of their societies – a careful viewer can often still distinguish higher-income residential areas from lower-income ones. For example, even in densely settled areas where people live predominantly in high-rise buildings, landscape cues signal whether the buildings cater more to high-income residents. Higher-income areas might have more balconies. Or, they might have greater security around building access; a security guard, for example, rather than security cameras, which may be prevalent in low- as well as high-income areas. Higher-income neighborhoods tend to have more restaurants – and more variety of dining options – and other services, such as clothing shops as well as corner groceries (the latter tend to cater to all residential areas). Buildings in higher-income areas also tend to have more consistently maintained landscaping around them. In low-rise areas, more expensive single-family homes tend to have larger yards around them – although this feature depends on the overall densities of the area. In some places, privacy fences indicate that a household has more disposable income, or the presence of household staff, such as landscapers, gardeners, and the like, signifies greater wealth.

While it might seem as though these visible differences in the landscape merely display objective differences in wealth, there are also ways that land-use law in many countries codifies separation of people by income but also by socially constituted categories such as race, ethnicity, religion, or caste. In the United States, for example, federal housing programs developed in the 1930s and 1940s to respond to the tremendous economic difficulties of the Great Depression built upon the racially biased logics about value that were embedded in the new urban theories of the time (Jackson 1985) (see box, "Race and Early American Urban Theory"). Specifically, in creating mortgage-stabilizing and funding programs, newly formed federal agencies drew on theories about racial and ethnic groups to assume that centered on its environmental and pro-density bona fides. In US cities with more typical patterns of human diversity (e.g., Oakland, Baltimore, or New York), pro-urban policies nearly always incorporate a politics of racial pluralism; in Portland, politics have until recently been largely silent about the path of Black lives in the city.

Who frames what and why?

These two place frames – ecomodernist icon and racial erasure – are of course not the only two frames in play for the region. Examples of others include a view of Portland as the pavement-ruined sacrifice zone at the northern edge of the fertile Willamette Valley; as a late-hippie funkadelic paradise of front-yard food gardens and locally grown artisanal marijuana; and, alongside Seattle, WA, and Vancouver, BC, as part of a triad of cosmopolitan urbanized outposts in the binational ecoregion of Cascadia. However, comparing the two place frames on which we focus here emphasizes two things: first, the power of hegemony to obscure or overwrite place counter-narratives; and second, how Blackness has been rendered more visible in Portland since 2020 through urban activism.

The ecomodernist icon frame is unambiguously politically left-of-center in the wider American context. In this framing, the story of the passage of SB100 is an epochal marker in the historical record, launching a new age in which Portland inexorably presses onward toward a greener, more sustainable future through state-led densification and a slowly intensifying culture of care.

A useful way to think about the reason that Portland's place politics have played out as they have is to ask: who benefits from the widespread adoption of a particular place frame at the expense of others? Portland's green/left identity has now drawn several decades of young talent to the region. This place frame has been harnessed to draw real-estate capital to putatively "green" development projects and infrastructure investments from state and federal governments.

neighborhoods with Black, Indigenous, or other People of Color in them would inevitably decline in value.

A national appraisal system evolved from these assumptions, where neighborhoods with Black residents were shaded red on appraisal maps. Banks were reluctant to fund mortgages in predominantly Black and Latino neighborhoods for decades, a process known as "redlining," because these areas were expected to lose value over time – due in part to the fact that these areas were overcrowded because racism prevented their residents from finding housing in other neighborhoods (Jackson 1985; Sugrue 2005 [1996]). The appraisal system helped to fund a post-World War II housing boom, underlain by the twin assumptions that new housing and white neighborhoods would always increase in value. As a result, and spurred by federal investments in highway transportation, white suburbanization characterized a significant portion of urban growth over the 1940s–1970s period. Although these patently racist housing financing policies were formally abandoned in 1968 with the passage of the Fair Housing Act, the years of disinvestment and limited housing options for Black and Brown Americans had set patterns in urban areas for years to come, with disinvestment – and concomitant decline – in inner-city Black areas. The idea that a neighborhood's social, ethnic, and racial characteristics shaped its economic value was by that time well entrenched, particularly in the white American imaginary, and was reinforced by the patterns of decline that occurred (Massey and Denton 1993; Duneier 2016).

Redlining may be particularly American, but the racist ideas underlying it are not. Many cities worldwide have housing markets that differentiate residents and districts not simply by wealth, or ability to pay, but by socially constituted categories of difference, be they race, ethnicity, caste, or religion. These patterns of social segregation are, in part, legacies of theories that connected race and ethnicity to spatial location and value, and are often institutionalized in financial decision making and urban planning.

See also related boxes "Urban Economic Processes" in chapter 2; "Race in Early American Urban Theory" this chapter; and "Landscapes and Power" in chapter 7.

The ecomodernist Portland place frame is thus useful to those invested in the conventional success of Portland in at least two different ways. In an explicit sense, it draws resources and attracts people who help the city-region to "win" in comparison with other US cities. But it also does work in rendering older and more unpleasant histories invisible. Paying attention to those older place narratives might provoke uncomfortable self-reflection for residents; it might also motivate a withdrawal of financial resources by external actors.

This explicitly left-leaning story narratively occludes a more problematic place frame: one of racial erasure. It is useful to highlight that the racial erasure place frame is not a critique from outside: white settler colonialists in Oregon actively articulated it for over a century. The paired cataclysms of the Great Depression and World War II serve (in this as in other political arenas) as a kind of historical reset mechanism, a new backstop against which a briefer history can be told without significant reference to the city that came before. Advocates of the ecomodernist icon frame tell the story of Portland as though the colonial period was prehistory rather than history, and as though the lack of Brown and Black bodies within Portland in the 1950s and 1960s was natural and largely apolitical, rather than actively produced through exclusion.

The racial erasure place frame is made clearer if one takes a significantly longer historical view. By the early 1970s, many of the relevant acts of explicit racial erasure seemed safely historically entrenched, leaving significant patterns of exclusion on the landscape. The character of Portland as disproportionately white in 1995, both demographically and culturally, required little explicit action from residents in the UGB era.

This whiteness was *specifically* anti-Black. The early white settler colonists of Oregon held many kinds of racial animus, including against the native people who were much more populous than people of African descent in those early days. But it was Black people in particular that were the most explicit targets of violence and exclusion during the period from early white settler colonialism until the present day.

Race in Early American Urban Theory: The Chicago School and W. E. B. Du Bois

Urban scholarly history tends to highlight and center the "Chicago School" of sociology as part of the origin story of urban studies. Scholars such as Robert Park, Ernest Burgess, Louis Wirth, and others at the University of Chicago in the 1910s through the 1940s used their urban location to observe and study communities and neighborhoods and draw conclusions about urban change and urban spatial patterns (Park, Burgess, and McKenzie 1925). These scholars sought to standardize knowledge of cities, and in doing so articulated theories that were innovative because they insisted on an understanding of urban neighborhoods as social areas with internal relationships and organization. But their theoretical explanations for urban spatial patterns also drew on pervasive social and racial biases in the United States at the time.

In particular, these scholars understood "race" as a biological category, rather than as a socially constructed category which projected social and economic significance on just one aspect of physical appearance. They understood neighborhoods as "natural areas," with the behavior of residents in an area being attributed to the prevailing race or ethnicity of those residents. Chicago School scholars were influenced by Darwinian theories about species competition and adaptation. They described urban spatial patterns in terms of the "succession" of different immigrant groups as people settled first in poor areas, and then some were able to move into more affluent areas as they became more economically stable. For Chicago School urban theory, successful groups were those that assimilated into dominant society, and those who did not were simply less skilled or adept groups (by virtue of their race and/or ethnicity). These theorists simply did not account for interpersonal or structural racism, economic processes, or relationships; nor were they troubled by generalizing from spatial patterns of ethnic settlement to characteristics of social groups. They did not explain why social categories of race and ethnicity evolved and carried social meaning, why people identified by

White settlers in Oregon were part of a much wider project of settler-colonial nation-building in the American West: that is to say, they were framing the United States in its entirety as a racialized place through the same kinds of actions that framed Oregon and Portland as places of white supremacy and domination. Yet Portland's anti-Blackness from the 1990s onward was active and visible in a way that was not true in other purportedly left-wing urban areas of the United States. This is presumably in part because most other major coastal cities had significantly larger minoritized and specifically Black populations. But the institutions of racism were also especially prevalent in Portland in the 1990s and 2000s. Volksfront, a local white separatist group founded in 1994, grew rapidly while advocating for a Pacific Northwest white ethnostate centered on the Portland area.

It is likely that many white residents of Portland in the 1990s and 2000s – particularly residents new to the region who were attracted by the growth boundary and the formal state commitments to mass transit – were oblivious to large parts of this history. It is typical and unremarkable that foundational horrors (like the treatment of native people in the United States) are often rendered obliquely in the histories that are preserved. It is banal to say that history is written by the winners. Yet, as noted above, significant minorities of residents voted to preserve explicitly racist text in the state constitution up to contemporary times. The social landscape of Portland, an urban region clearly racialized as white, was not remade when the city's residents began framing their story differently. Instead, national events served to challenge the frame of racial erasure in ways that are still emerging and being contested.

The aftermath of the murder of George Floyd and the re-racialization of Portland's dominant place frame

In the summer of 2020 and in the midst of the Covid-19 pandemic, a Black man named George Floyd was killed by those categories settled where they did, or question whether landlords and real-estate agents discriminated against some people because of their race or ethnicity.

Urban theorist W. E. B. Du Bois, working in Philadelphia in the late 1890s, also wanted to understand the social diversity and segregation of different groups in cities (Du Bois 2014 [1899]). Du Bois – who was Black, and in 1895 had earned the first PhD that Harvard University had granted to a Black person – pointed to social structures and biases, not cultural groups or characteristics, that shaped or limited opportunities for different groups of people in urban areas. He focused on the predominantly Black neighborhood of the Seventh Ward in Philadelphia, examining how people were treated. Most Black residents in Philadelphia (and indeed, throughout the United States), regardless of skills and education, had very limited opportunities for work, usually in domestic service. Du Bois also documented that even the relatively well-to-do (Black) residents of the Seventh Ward could not live elsewhere in the city because no one would rent or sell housing to them in any neighborhood except the Seventh Ward. (Thomas Sugrue [2005 (1996)] offers strikingly similar evidence for Detroit in the period both before and after World War II, demonstrating a broader social dynamic of racism at work.)

Despite the fact that Du Bois wrote earlier than the Chicago School scholars, his work was not taken up by these or other scholars until many years later (Morris 2015). Du Bois worked for many years as a professor at a Black-serving institution (Atlanta University), while his work went largely unrecognized by white scholars. The Chicago School scholars, while training a diverse set of students, had the advantage of their whiteness in a white-dominant society. Their work was tremendously influential, even outside of the United States, leading to the development of urban theory that unproblematically categorized people by race and ethnicity and associated them with particular land uses and building conditions, largely ignoring structural political and economic forces, and the role of racism, that produced social categories of difference as meaningful.

See also related boxes "Residential Segregation and Redlining in America" this chapter; "Landscapes and Power" in chapter 7.

A Black Lives Matter street protest[2]

Black Geographies

Scholars studying cities (including anthropologists, sociologists, geographers, and planners) have long wrestled with race as an important category of analysis. However, notwithstanding early work by Black scholars including W. E. B. Du Bois (see box, "Race in Early American Urban Theory"), these urban disciplines have long been dominated by white scholars pursuing racial analysis from their own embodied perspective. While there is a long line of urban scholarship critical of segregation and discrimination (Massey and Denton 1993; Sugrue 2005; DeFilippis and Wyly 2008; Omi and Winant 2014), the perspectives of Black urban residents themselves have often been overwhelmed in urban scholarship by quantitative analyses of social or spatial patterns of racialization (see Clark 1991).

In the last two decades, a scholarly subdiscipline called Black geographies has emerged from geography whose participants seek to change this dynamic. They seek both to center Black lived experiences in their scholarship and to highlight how the perspectives of Black researchers shape the kinds of questions that are asked. Fundamentally, scholars in this area highlight the ways that spatial exclusions are constituted through institutionalized racism, and that place-making nonetheless persists and occurs from the "margins" through Black agency.

Scholars working in the Black geographies tradition have written from diverse perspectives about place. Katherine McKittrick (2006; McKittrick and Woods 2007) works to excavate the spatialities of Black lived experience. Adam Bledsoe and Willie Wright (2019) describe the plurality of Black geographical experiences. Other scholars build specifically on a relational place-making approach to argue for the ways that Black people appropriate and deploy elements in the urban landscape that were built in service of racist societies to make meaning and places for their communities (Allen 2020; Allen, Lawhon, and Pierce 2019).

Scholars writing in other disciplines, such as sociology and history, have also worked to center Black and Brown police officers in Minneapolis, United States. A Minneapolis Police Department officer, Derek Chauvin, knelt on the neck of Floyd for more than nine minutes as he lay still, complaining of difficulty breathing and fear for his life. After Floyd became unconscious, Chauvin continued to pin him with his knee as he died.

A series of intense protests began across the country that lasted into the following year. Like most mass protests, those in Portland had a clear trigger coupled to a more complex and rooted set of motivations. While protesters were angry about Floyd's death, they also mourned the many other Black Americans whose lives had been cut short by white supremacy and (specifically) police brutality.

While Portland is half a continent away from Minneapolis, the killing struck a chord for local residents and particularly Black residents. More than 100 consecutive days of protests followed, with crowds ranging from the hundreds to the thousands. Most of these protests were peaceful; some involved tense interactions with local and federal police; a few became violent. Even after air quality from nearby forest fires cut the streak short, regular protests continued into the following year.

In one sense, the protesters in Portland were participating in a kind of national vigil in honor of Floyd and the many other Black people who have been killed by police. They advocated for defunding police departments and replacing their functions with social services agencies. They memorialized Kendra James, Quanice Hayes, and others who have been killed by the Portland Police Bureau.

But they also used the moment of national attention to reframe Portland through its longer history of racism. Protesters highlighted the *longue durée* of racial abuses in Portland specifically, as opposed to the more general history of American racism to which Oregon is a party. The rawness of this explicit lineage of anti-Blackness was emphasized when then-President of the United States Trump ordered federal police into Portland, nominally to protect federal assets within the city. While the move was applauded by national conservatives, for Portland's activists

> life experiences and agencies in their scholarship. Marcus Anthony Hunter (2013) and Zandria Robinson (2014), working together (2018) and separately, have focused on a Black American urban sense of place. Historian Eric Avila (2014) has highlighted how Black and Latino/a/x communities in the United States reconfigured their place identities in the wake of neighborhood loss from twentieth-century highway construction.
>
> To date, the Black geographies project has largely been centered on the experiences of the descendants of slaves in North America, and to a lesser degree the experiences of immigrants and descendants of slaves in Europe. While there is also scholarship on the role of Black identities in postcolonial contexts (Simone 2009), at time of press the Black geographies and Postcolonial Urbanisms projects remain mostly in parallel rather than mutually integrated (Noxolo 2022).
>
> *See also related box "Landscapes and Power" in chapter 7.*

the action re-emphasized that Portland was especially sensitive to claims that Black lives matter. Federal officers weren't deployed in most cities in the aftermath of the Floyd killing, but they were deployed in this very white one.

Young white left-of-center residents were an important part of the Floyd protests in Portland. This series of events was the first moment in Portland's history when white people participated in sustained public protest against anti-Black violence. Portland's population is still very white, but the framing of the city has shifted. The whiteness of the ecomodernist icon frame is much more visible at the time of writing, in early 2022, than it was even five years earlier. These events highlight the ways that seemingly dominant and unquestioned frames come to be exposed as incomplete and inadequate. The current struggles in Portland over its identity and setting as a site for all of its residents show the many ways that people can actively shape place frames, and the possibilities for social and political engagement around urban meaning and experiences.

This chapter might help you to think about cities like Amsterdam, Netherlands; Zürich, Switzerland; Cape Town, South Africa; and Savannah, GA, USA.

–6–
Chongqing: International Cyberpunk Marvel/National Policy Innovator

Amy Y. Zhang

Chongqing is located in the southwestern region of China with a population of 32 million in 2020. During the Second Sino–Japanese War (1937–1945) and continuing until the end of the Chinese Civil War in 1949, it served as the wartime capital for the Republic of China administration. After the establishment of the People's Republic of China in 1949, Chongqing was a municipality in Sichuan Province. Then in 1997 the central government of China separated Chongqing from Sichuan Province to establish it as a direct-administered municipality, joining Beijing, Tianjin, and Shanghai as the only four municipalities that are of the same rank as provinces in China.

In China, a direct-administered municipality is more like a province of smaller size and is not strictly a "city" in the conventional sense because it typically consists of a central urban area and surrounding rural areas under the same administrative unit. The urban area is divided into districts and the rural areas contain county-level divisions, with subdivisions that are further defined based on townships and villages. With the increasing expansion of urban areas over the years, some rural counties have been reclassified as districts. But these new districts do not necessarily resemble their highly urbanized counterparts, and many contain a mix of urbanized built-up areas and rural land use. Sometimes, the reclassification of rural counties into urban districts is not the result of urbanization but is instead a means for facilitating ongoing urban expansion. Hence, despite Chongqing having a land area of 82,403 sq. km in 2019, its built-up area was just 1,680.52 sq. km, or 2 percent of the overall land area.

Due to its location at the transitional area between the western region and the central plain of China, Chongqing has long been designated with strategic importance by the central government even before it acquired its status as a direct-administered municipality. As a result of both its geopolitical strategic importance and inland location, the economy of Chongqing was dominated by military-related and heavy manufacturing industries, such as motor vehicle production and iron and steel production, until the early 2000s. This strategic importance is further reflected in the change of Chongqing's status to a direct-administered municipality in 1997, which was also partly due to the important role it played in coordinating the relocation of people displaced by the construction of the Three Gorges Dam.

As the only direct-administered municipality in the western region, Chongqing was positioned to spearhead China's Open Up the West program when it was initiated in 1999. The Open Up the West program, or Great Western Development Strategy, is the central government's response to the fact that economic development in the western region was lagging behind the eastern coastal region. Chongqing's role in the Open Up the West program has benefited it with infrastructural development, and also gives it opportunities to conduct policy experiments ahead of the rest of the region. Particularly, in 2010, the Liangjiang New Area was established in Chongqing as only the third state-level "nationally strategic new area" in China and the first of its kind in the western region at the time, following the Pudong New Area in Shanghai and the Binhai

New Area in Tianjin. These state-level new areas are delegated the power to carry out experimental policies and reforms on issues that are deemed to be nationally significant and therefore receive preferential policies and privileges directly from the State Council, which gives them notable advantages to attract business and investment (Kean 2019). These new areas tend to be located at the periphery of urban areas to provide spaces for new development. The Liangjiang New Area takes up 1,200 sq. km of land (with 550 sq. km available for construction) and covers parts of three urban districts that are to the north of the center of Chongqing, superimposing on existing administrative divisions of the city. It is identified by the State Council as a pilot area for urban–rural integration strategy as well as to play an important role in "opening up" the western region and developing agglomeration economies (Martinez and Cartier 2017).

The establishment of the Liangjiang New Area in Chongqing has facilitated the transformation of the municipality's economy away from the weapon and heavy manufacturing industries. After the 2008 global financial crisis, the municipal government of Chongqing had already been making deliberate efforts to attract computing notebook manufacturers that were considering leaving China's more developed eastern coast. Through a combination of providing land and labor at a comparatively lower cost, investment in infrastructure, and preferential policies in the New Area, such as tax breaks, major companies including Hewlett Packard and Foxconn were persuaded to relocate to Chongqing. Chongqing consequently became the "land of laptops." In 2014, 40 percent of the world's laptop computers were produced there (Yang 2017: 26).

Thus, since the late 1990s, Chongqing has played the role of leader in China's western region, and its policies, strategies, and development are seen as representing a kind of future for the rest of the region. This view of Chongqing as representing a particular kind of future is also reflected in the two place frames discussed in this chapter. What follows first examines how a focus on Chongqing's built environment leads to a place frame that characterizes Chongqing as a *futuristic, cyberpunk city* on the global scale. Attention then shifts to a different place frame that stems from the municipal government's policy experiments. This place frame positions Chongqing as the place that provides *a solution to the problem of inequity* in Chinese cities and promises a more egalitarian mode of urban development.

Furthermore, as shown in this brief introduction to the city, Chongqing's development, like that of any other city, has been shaped by both national and global contexts. And the municipal government has tried to take advantage of particular dynamics to (re)define the position of the city within these contexts and control the direction of its development, such as the efforts made in economic transformations post-2008. The two place frames discussed here also reflect the ubiquitous interactions between the broader context and dynamics that Chongqing is situated in and its municipal government's attempt to shape the city into an economically strong, well-known player in the global arena as well as within China.

As the rest of this chapter will describe, the first place frame is shaped by an existing global trend of associating Asian cities with "the future," especially in the western imaginaries. While this global dynamic and its influence on framing the city are beyond the municipal government's control, it has embraced this framing and capitalized on it for enhancing the image and brand of Chongqing to attract more attention to the city. In contrast, the second place frame is a claim made by the municipal government of Chongqing. But, similarly, it is also shaped by a broader context, namely, a national context of widespread inequalities in cities, faced especially by rural migrant workers. This framing is then more politically significant against this context and helps the municipal government to assert a leading role for Chongqing in China.

Chongqing as a city with a weird, cyberpunk, and futuristic built environment

Many cities across the world try to curate unique, interesting, and memorable images for themselves hoping to attract attention and, subsequently, investment (see box, "City Branding"). Finding a distinctive niche becomes necessary. Among Chongqing's efforts to establish a brand for the city, cultivating a level of "strangeness" in its urban landscape seems to have been a successful one. For example, the now-closed Foreigners' Street, which was a 24-hour free-entry amusement park opened in 2006, was known for its bizarre architecture, including "the largest toilet"; and Huangjueping Main Street, near the old campus of Sichuan Fine Arts Institute, was painted in 2005 with colorful cartoons and shapes covering whole buildings along the street to turn it into a "graffiti street." However, what effectively establishes this place frame and brand for Chongqing is not its purposely created segments of bizarre landscape. Instead, the place frame is built through an online social media circulation of the imagery of the city's mundane residential buildings and public transit infrastructure, which were produced through the interactions between Chongqing's topography and its urban development history.

For people who live in or are from the nearby region, Chongqing is known for the verticality of its built environment, which partly results from its topography. Chongqing is nicknamed "Mountain City" as its central urban area is built on mountainous terrain, with two mountains each defining its eastern and western boundaries, while the Yangtze and Jialing rivers cut across the city. Its traditional urban core is a narrow and steep peninsula, Yuzhong, which is bordered by the two rivers.

This topography means that urban life in Chongqing has long been organized through multiple vertical levels. Its urban core has traditionally been divided into an "upper-half city" on the mountain top and a "lower-half city" halfway up the mountain, which are connected through steep staircases and cable cars. This vertical division also translates into a hierarchy. Although the city originated in the "lower-half city" (sometimes called the "mother city" by local residents), during the time when Chongqing was the wartime capital, the "upper-half city" was designated to be where the political and economic power would locate and the "lower-half city" became mainly for residential uses. This division and hierarchy continue today: the "upper-half city" saw more development after economic reform and is where the central business district (CBD) sits, whereas the "lower-half city" was marginalized and overlooked in urban development plans for years until it became the focus for urban renewal and historical conservation starting in 2015.

While Chongqing's urban landscape was already defined by its verticality, a more striking vertical-built environment has been produced in the city from a real-estate construction boom started in the 1990s, especially following its status change to a direct-administered municipality and the Open Up the West program. The construction boom was mainly characterized by high-rise buildings, both apartment buildings and skyscrapers. Many of the newly built high-rise apartment buildings traverse multiple vertical levels defined by existing roads and streets that are carved into the city's mountainous terrain. These buildings also often utilize the existing roads and streets to create vertical connections, such as having multiple entrances to a building on different floors according to the streets on different vertical levels that the building is built against.

One example is the White Elephant (Baixiang) Street residential tower block, which consists of six 24-story apartment buildings. These buildings are connected through bridges and have entrances at three levels, on the first, tenth, and fifteenth floors of the towers. The vertical connection in this case was not only a result of taking advantage of existing vertical divisions and organization of the urban landscape, but was also born out of necessity as none of the six 24-story buildings have elevators. The lack of elevators has been attributed to a combination of factors that were present when the tower block was constructed from 1983 to 1992, including relaxed construction regulations at the time, the relatively high cost of elevators, and the developer's incentive to maximize profits. The verticality of Chongqing's urban landscape makes it possible to have entrances on three different floors of the buildings and makes living in a high-rise apartment tower without an elevator slightly more bearable.

City Branding

The practice of associating cities with images and identities has long existed; nicknames such as "The Big Apple" (New York City) and "The City of Love" (Paris) are widely known and firmly connected with their corresponding cities. People form their understanding, knowledge, and perception of a city by both directly experiencing the city and encountering the image and identity of the city communicated through media representations and its built environment (Holloway and Hubbard 2001). Such images and identities may be deliberately established: for example, "The Big Apple" became synonymous with New York City through a tourism campaign in the 1970s. They may be outcomes of an industry dominating a city's economy, such as Los Angeles in relation to the film industry. Detroit's historic role in the American automotive industry gave it the nickname "Motor City." They may also be formed in more collective and grassroots ways (Chongqing's nickname "Mountain City" is an example) or through popular culture.

The images and identities that cities are associated with can help form "a sense of sociocultural 'belonging'" (Evans 2003: 421) for residents, and they can also be leveraged as "brands" for cities to differentiate themselves, especially from their competitors. Arguably, cities have been trying to make themselves more appealing for as long as they have been competing with each other for capital, labor, and trade (Ward 1998). However, city branding as a specific urban development strategy entered a new stage in the 1990s as part of a shift in the mode of urban governance from a more managerial stance to a more entrepreneurial one since the 1970s (Harvey 1989a). Harvey (1989a) argues that as cities face more intense inter-city competition for attracting increasingly mobile capital across the globe, city governments move away from managerial functions, such as providing services and benefits to local residents, toward more entrepreneurial actions, seeking to improve their cities' competitiveness. The attractiveness of a city to potential

Verticality of Chongqing[1]

investment heavily relies on the image of the city, which makes branding an important component of urban entrepreneurialism (Short et al. 1993).

Similar to branding a consumer product, city branding has been done through creating logos and slogans. But "re-imagineering" a city's built environment and staging activities like festivals and mega-events are equally significant for city branding (Kavaratzis and Ashworth 2005). The emphasis is on creating a look and feel for the city in accord with widely agreed aesthetic qualities of a "world-class" city (Harris 2011; Aiello 2021) to make it "appear as an innovative, exciting, creative, and safe place" (Harvey 1989a: 9). Existing research focuses especially on the use of cultural and artistic elements by postindustrial cities for changing their images and economic structure (Evans 2003). This trend is seen as an effect of the perceived success of Bilbao, which managed to turn away from economic decline in the 1980s toward regeneration and a new economy based on tourism and the service industry after the building of the Guggenheim Museum with a spectacular design by Frank Gehry in 1997 (Gomez 1998). But even for cities that have long been global financial centers, such as London, "re-imagineering" and rebranding become necessary when new competitors emerge. In response to the perceived challenge from Frankfurt to London's financial center status in the late 1990s, a new generation of spectacular architecture, starting with the Swiss-Re Tower (nicknamed "the Gherkin"), has been added to London's skyline since the early 2000s to signify the city's preferred new image of being a playground for footloose transnational capitalist class as opposed to its old, English-centered, image (Kaika 2010).

City branding could be used for constructing a strong collective identity if the images it creates and the stories it tells are well connected with the place (Zukin 1995; Kavaratzis and Ashworth 2005). But when used primarily for competitive purposes, city-branding practices tend to follow a handful of assumed successful formulas, such as the attempts to replicate the case of Bilbao by other

Although people in Chongqing have been living with its vertical landscape for a long time, Chongqing's built environment started attracting increasing attention from a wider audience around 2016. The interest is largely due to the circulation of the imagery of Chongqing through social media platforms, first within China and then internationally. Tourists across China came to the city to see its verticality firsthand and document and share their amazement at the built environment with photographs, videos, and social media posts, which generate even more interest in the city.

In addition to the White Elephant Street tower block, another site that is often documented and shared online by visitors is the Liziba station on Chongqing's number 2 light-rail line. Here, the intersection between a construction boom under relaxed regulations and the city's need to accommodate its expanding size led to a planning compromise that incorporates the station into a 19-story apartment building built on the route of the light-rail line. Since the light rail is a monorail, the station is in the middle of the apartment building and occupies the sixth to eighth floors, creating a rare scene of trains passing through an apartment building. While the station has been open since 2004, it had not been viewed as a special feature of the city until around 2017 when growing numbers of tourists stopped by to take photos and videos of trains passing through the building. There are even Google reviews of the station, dated back as early as 2018 and in multiple languages, lauding the "unique," "magic," and "marvelous" "wonder" that one can observe there (see image below).

Obviously, Chongqing is not unique in having a vertically organized urban landscape. Hong Kong, for example, has a quite similar built environment that is defined by multiple vertical levels and vertical connections such as the Central–Mid-Level Escalator and walkway system. However, the built environment of Chongqing presents a level of "extremeness" and by extension a certain "weirdness," exemplified by the two sites discussed here, which draws special attention and evokes amazement. Gradually, a place frame of Chongqing has been constructed through social media and general

> postindustrial cities. This kind of branding often has little connection with existing local identities or may be done through manipulating local history and culture, therefore denying or excluding significant aspects of the existing identity of the city (Evans 2003; Harris 2011). Lacking local connections also raises doubts about the sustainability of the brand, and cities adopting similar branding strategies create sameness and dilute the effect of branding (Griffiths 1998; Evans 2003). City branding, therefore, is speculative and carries high risk. Not only does it divert both attention and resources away from addressing local needs and socio-economic inequality, but city branding also does not always deliver the benefits city governments claim it will.
>
> *See also related boxes "Urban Economic Processes" in chapter 2; "The Growth Coalition" in chapter 3; and "Landscapes and Power" in chapter 7.*

media coverage of the city, which characterizes Chongqing as a city with a weird and even cyberpunk-esque built environment.

As indicated above, the circulation of the imagery of Chongqing's vertical landscape occurs both within China and globally, especially in Anglophone social media and media spaces. Although a similar focus on the "weirdness" of Chongqing's built environment also exists in Anglophone commentaries, the visuals circulated in Anglophone media spaces especially emphasize the size, scale, and height of the built environment; the contrast between striking skyscrapers in the CBD and the decay of the "lower-half city"; and nighttime views that add an "otherworldliness" to the urban landscape through neon lights. Built upon these representations, the place frame of Chongqing that appears in Anglophone media spaces characterizes it more as a city with a cyberpunk and futuristic look: a "cyberpunk urban jungle" (Roast 2021: 82). (Cyberpunk is a subgenre of science fiction focusing on a dystopian future of oppressive society dominated by advanced computer technology.)

Liziba station[2]

This characterization of Chongqing is the continuation of an existing place frame in western discourse where "the future" is imagined and visually presented to be somewhat Asian (Asian futurism), and oftentimes East Asian in particular (Borgonjon 2018) (see box, "Asian Futurism, Sinofuturism, and Orientalism"). This trend is especially noticeable in western science-fiction movies, ranging from the geisha holographic advertisement in *Blade Runner* (1982) to Shanghai's urban landscape acting as the backdrop for *Her* (2013). The video essay *Virtually Asian* by the artist Astria Suparak (2021) shows a collection of such examples, drawn from major western sci-fi movies spanning four decades. Suparak's video essay also criticizes the curious absence of Asian characters and their agency in these screen productions that are set against a hollow Asian aesthetic and backdrop, echoing other cultural critics (Chan 2016; Borgonjon 2018). The ways in which Chongqing's urban landscape is visually represented in Anglophone media spaces, as identified above, have a strong association with the typical cyberpunk-esque Asian aesthetic employed in these sci-fi movies. And just like the anonymous Asian cities acting as backdrops in western sci-fi movies, the visual representations of Chongqing's urban built environment that form the basis of the place frame are almost always devoid of human beings. This visual similarity between the imagery of Chongqing circulated in Anglophone online spaces and the urban Asian aesthetic in western sci-fi movies may be seen as both the basis for the place frame and the effect of this existing trend. This place frame of Chongqing also helps reinforce these futuristic and cyberpunk-esque depictions and imaginations of Asian cities.

The place frame of Chongqing and Asian futurism more broadly can also be seen as part of a continuity in depicting cities located outside of the Euro-American center as fundamentally different from western cities. The essentialist and reductionist characterizations of non-western and Global South cities have been criticized by urban researchers since the early 2000s, especially scholarship that equates these cities with "underdevelopment," mega-cities (Robinson 2002), and slums (Roy 2011). However, reductionist representations of cities persist. For some Asian cities, technology, speed, and scale are now the new registers that set them apart from their western counterparts. Asian futurism and the place frame of Chongqing are examples of how some Asian cities are depicted as incomparable with western cities.

Despite the unexpectedness of and lack of control from the local level over this place frame's initial production, the municipal government of Chongqing has been quick to capitalize on the framing. First a viewing platform next to the Liziba station was opened in August 2018 by the city to accommodate the amount of tourists that the spot attracts, practically turning it into an officially recognized tourist attraction. More recently, Chongqing announced in May 2021 that a second scene displaying "trains passing through a building," and thus another tourist attraction, would be created at the future Hualongqiao station on the number 9 light rail, which is expected to start operating in 2022. This station would be incorporated into the future Chongqing International Trade and Commerce Center located in the International Business District on the Yuzhong peninsula, the construction of which is expected to be completed in 2022. The fact that the 99-story, 458-meter-tall Chongqing International Trade and Commerce Center, upon completion, would become the tallest building in Chongqing only adds to the spectacle that this plan intends to create. This intention is made more obvious with an open call for concept design proposals for the exterior of the station to enhance the image of the site and its special feature, "trains passing through the building." Thus, while the place frame is constructed by many different visitors to the city and social media posts about the city, it conveniently provides the ingredients that the municipal government can use to build and enhance an image of the city as one that appears to be exciting, spectacular, and innovative.

What is missing in this place frame and the subsequent urban development strategy are the residents of the city, despite the fact that the basis of this place frame comes from their everyday environment. Most media coverage about the city's relatively new fame for having a "weird" and "special"

Asian Futurism, Sinofuturism, and Orientalism

Asian futurism refers to the tendency of imagining "the future" as "Asian" in western cultural products and discourse (Chan 2016; Borgonjon 2018). This trend emerged in the late 1980s, when Japan became a leading figure in developing industries regarded as "the technologies of the future – with screens, networks, cybernetics, robotics, artificial intelligence, simulation" (Morley and Robins 1995: 168). As technology is viewed as central for the future, the future, by this logic, becomes Japanese. This trend of locating "the future" in Asia continued with the rise of the "Four Asian Tigers" (specifically, rapid economic growth through the 1990s in South Korea, Singapore, Taiwan, and Hong Kong) and then China (Roh, Huang, and Niu 2015). The fast-paced modernization strategy pushed by the Chinese state, growth of Chinese technology industries, and drastic changes to the landscapes of large Chinese cities from the 2000s also fueled a subgenre of Asian futurism in recent years, namely Sinofuturism. In this discourse, China is presented to be already in the future in comparison with the West, evidenced in particular by the skylines of its ever-growing large cities (de Seta 2020). Chongqing, therefore, following Shanghai and Shenzhen, becomes the latest choice as the representation of a future in western commentaries (Roast 2021).

Asian futurism, as a discourse constructed by the West about Asia, has often been discussed in relation to the previous western discourse about the "East" or "Orient" as critiqued in Edward Said's groundbreaking book *Orientalism* (1979). The premise of Said's argument is that, "as both geographical and cultural entities," both the East ("the Orient") and the West ("the Occident") are "man-made" (1979: 5); namely, they come into existence and are sustained through discourses and imageries about them. Said examines how "the Orient" is depicted and represented in western knowledge as a backward, passive, eccentric, and inferior Other –xconstructed as the opposite of the West. Orientalism, therefore, refers to this "manner of regularized (or Orientalized) writing, vision, .and study, built environment has little to no mention of residents' reactions. Online commentaries show that Chongqing's residents are quite fond of the city's particular topography and built environment, but they are more ambivalent about the "magical" and "surreal" labels being used by tourists to characterize the city. On one rare occasion, a resident is quoted in a news article about the popularity of the White Elephant Street tower block, commenting that while many visitors view the tower block as something special, to those who live there "it is not a big deal; we got used to it long ago" (Lang 2018).

Not only are they missing from the narrative, local residents also do not benefit much from the impacts of the place frame. In fact, the viral popularity of the Liziba station has already become a nuisance for nearby residents and commuters who ride the light rail regularly. The municipal government's choice to open a viewing platform encourages even more tourists to visit the station, crowding the station and nearby streets. The viewing platform also limits the angle from which photos can be taken, and some tourists have taken to knocking on apartment doors in nearby residential buildings, hoping to find more options for photographing. Annoyed by these unwelcome visitors, nearby residential blocks had to post notices with the phrase "Private residence – no visitors" to deter tourists. The plan to recreate the scene of "trains passing through a building" at the future Hualongqiao station in the center of Chongqing's International Business District further demonstrates the municipal government's focus on image creation that has little to do with the day-to-day life of an ordinary resident of Chongqing.

Chongqing as a place that promises a more egalitarian mode of urban development

In contrast to the absence of local residents in the previous place frame, this second place frame is based on attention to residents of the city. In particular, this place frame of Chongqing is largely a claim made by the municipal dominated by imperatives, perspectives, and ideological biases ostensibly suited to the Orient" (Said 1979: 202). The purpose of this discursive construction of "the Orient" is to establish and reinforce the West's position as the center of modernity and subsequently its power and domination over "the Orient." Orientalism is a product of and serves western hegemony by "reiterating European superiority over Oriental backwardness, usually overriding the possibility that a more independent, or more skeptical, thinker might have had different views on the matter" (Said 1979: 7).

Although Asian futurism, which projects "the future" onto Asia, may be seen as the opposite of the Orientalist idea of equating the East with "backwardness," these two are not so different in essence. Scholars who examine Asian futurism and its subgenres point out that this trend reflects western anxiety facing technological and economic development in Asia on the one hand, and stagnation and even decline in parts of the West (Roh, Huang, and Niu 2015; Chan 2016; de Seta 2020; Roast 2021). The superiority of the West as the center of modernity, established through Orientalism, is called into question and threatened (Morley and Robins 1995). Asian futurism is therefore identified as a way for the West to reassert its centrality in this new context "by creating a collusive, futurized Asia" (Roh, Huang, and Niu 2015: 7). Jean Baudrillard's (1988: 79) characterization of Japan as "a satellite of the planet Earth" exemplifies how Japan is othered by such discourse: "Japan not only is located geographically, but also is projected chronologically" (Ueno 1996: 95). Scholars have coined terms such as "techno-Orientalism" (Morley and Robins 1995; Ueno 1996) and "high-tech Orientalism" (Chun 2006) to reflect that Asian futurist depictions of Asian countries conjure up new stereotypes of these places as being hypo- and hyper-technological and "Asiatic bodies functioning as gatekeepers, facilitators, and purveyors of technology" (Roh, Huang, and Niu 2015: 13). Thus there is a continuity between Asian futurism and the old Orientalist framing in othering, exoticizing, and essentializing Asia in western discourse, albeit through government about innovative social policies resulting from its reform of rural migrant workers' status in the city.

The reforms address China's household registration system. Household registration, or *hukou* as it is called in Chinese, is used in China to control population mobility. Everyone in China has a *hukou* registration that designates where they are supposed to live and should receive services (see box, "Urban Migration and *Hukou* in China"). Without a local *hukou*, one may not be able to access necessary social welfare, be eligible for home ownership, enroll children in local schools, and so on in the city where one actually works and lives; or one may have to pay a much higher cost (e.g., school tuition) or be a longer-term resident (e.g., having paid 5 years of tax in the same city) to access the same opportunities or have the same rights as residents with local *hukou*.

It can be extremely difficult to change a person's household registration, especially to gain one in large cities like Beijing and Shanghai. For example, some of the major avenues to gain a *hukou* in Beijing include joining central and municipal governmental agencies and public institutions as a civil servant, getting a job in companies and universities that have (limited) quotas for transferring *hukou*, or having lived in Beijing with a temporary residence permit and continuous social security contributions for seven years. To acquire a temporary residence permit in Beijing, a person must have already had a stable job and residence for six months in Beijing and be expected to keep both for another six months.

As a result, low-income rural migrant workers are much less likely to gain local *hukou* in the cities where they migrate as they tend to have unstable employment and residence, making them the "floating population" of China. Despite their essential labor roles in building and serving cities, rural migrant workers largely do not have urban status or the rights associated with status in those cities, inherently being treated there as second-class citizens. Rural migrant workers thus induce lower costs because both their labor tends to be valued less and their social reproduction requires less contribution from city governments and employers in comparison with those with local *hukou*. In fact, China's

> different terms. Also similar to Orientalism as originally defined by Edward Said, Asian futurism is an expression of western power, which denies Asian countries "the possibility of challenging and negotiating representation in the coeval present staked out by Western knowledge production" (de Seta 2020: 89).
>
> *See also related box "Modernism and Urban Design" in chapter 2.*

post-reform economic growth has been built upon this group of "super-exploitable, yet highly mobile and flexible industrial workers" (Chan and Wei 2019: 431).

In 2010, Chongqing launched the largest program in China to remove the barriers to rural migrant workers acquiring urban status, making them eligible for state welfare, healthcare, and education benefits in the city. This ambitious program was justified by the argument that it would make Chongqing more attractive to both employers and the labor that it needed for its growing electronics manufacturing sector. It is also tied to the status of the Liangjiang New Area as a pilot area for experimenting urban–rural integration strategies. Along with the *hukou* reform is the provision of affordable public rental housing by the municipal government, which pledged to construct 40 million sq. meters of such housing with rent set at lower than 60 percent of the market rate. The public rental housing estates are designated to meet the needs of migrant workers (both rural migrant workers and university graduates from outside of Chongqing), as well as of low-income local residents. There is no *hukou* restriction on the eligibility for public rental housing so migrant workers do not need to wait for the conversion of their statuses into local *hukou* before applying for public rental housing. However, high demand for such housing means that a lottery system is used to decide who can actually move into one.

The public rental housing program is possible because the municipal government of Chongqing had over the years banked large amounts of land instead of assigning it to private developers (in China, the state owns urban land and leases its use rights to developers) (see box, "Land Markets"). The control of land is crucial for the construction of public rental housing. It also enables the municipal government to fund its *hukou* reform program using profits generated from land appreciation since a greater number of urban *hukou* residents means higher levels of public spending to support those residents with social services (Huang 2011).

Recognizing that there is still a limit to the land that the municipal government has in its storage, a rural land exchange program has also been introduced, based on the notable size of the rural hinterland in Chongqing. This rural land exchange program's primary purpose is to make it possible for the municipal government to expropriate and bank rural land near the urban fringes for urban development. At the same time, it needs to maintain a minimum level of farmland in the region, as required by the central government. To achieve these goals, rural households located in less optimum locations for potential urban development are mobilized to restore their homestead for farming, after which a land certificate is assigned to each participating household. These certificates are then bundled together by Chongqing County Land Exchange to form an appealing size that both private developers and the municipal government can buy. Instead of the actual rural land that these certificates are attached to, the buyers would gain the use rights of equivalent sizes of rural land expropriated for urban development. The price, therefore, is based on the optimum location of the land that is permitted for development, which also means that the payment received by participating rural households (after deducting costs) can be much higher than the typical compensation from land acquisition (Zhang 2018). Rural households can then choose to continue farming on restored land or use the payment they receive to relocate to the city where the *hukou* reform also makes it possible for them to obtain urban status.

Together the urban *hukou* reform and rural land exchange program constituted the basis for the municipal government

Urban Migration and *Hukou* in China

Since the Industrial Revolution, cities have been seen as centers of economic and social activities that offer opportunities and diverse social expressions. People may migrate from rural areas to cities or between cities for more and/or better opportunities. Cities may also try to attract a particular workforce for developing different industries. Urban migration tends to be seen as a result of these economic and cultural dynamics. However, in China, urban migration has also been shaped by the household registration (*hukou*) system since the 1950s: in this system, every person is required to register in one and only one locale (cities, towns, villages, etc.) as their presumed regular residence, the only place where they are entitled to state-funded social programs, including formal housing and public schools. This registration is also hereditary, namely, a child's *hukou* is not determined by where they were born but by their parents' *hukou* (Chan and Zhang 1999; Johnson 2017).

The *hukou* system was introduced to cities in China in 1951 and to rural areas in 1955. Its main purpose at first was to monitor population mobility. However, at the time, China followed the Soviet Union in prioritizing industrialization in cities as an economic development strategy, using a centralized planned economy to channel resources and benefits toward cities. This strategy stimulated an influx of rural migrants to cities, creating economic burdens in those areas (Chan and Zhang 1999; Fan 2008; Chan and Wei 2019). The first set of national *hukou* legislation was then introduced in 1958 to curb this flow. During this era of planned economy until China's economic reform in 1978, internal migration was only possible when allowed by the state for the employment needs of state enterprises and governments.

The introduction of a market economy after China's economic reform initiated in 1978 induced changes to the pattern of urban migration in China as a non-state employment market came into existence, making living outside one's *hukou* locale possible as well as desirable in of Chongqing to make the claim that it can perpetuate economic growth without exacerbating urban–rural inequality: rural migrant workers can obtain urban status while rural households in remote areas can also reap financial benefits from urban expansion and development. A place frame of Chongqing has therefore emerged which depicts the city as a place that offers a solution to the urban–rural dual structure commonly found in large Chinese cities and as a place that promises a more egalitarian mode of urban development. At the beginning of the Chongqing reform, the Marxist geographer David Harvey (2012: 64) viewed it as "a purportedly radical shift away from market-based policies back onto a path of state-led socialist redistribution . . . to reduce the spiraling social inequalities that have arisen over the last two decades across China." But he also cautioned that the large scale of Chongqing's reform may "[accelerate] the dispossession of land from rural uses" (Harvey 2012: 64) and lead to discontent.

The characterization of Chongqing as representing a future direction for Chinese cities was first strengthened in 2010 when the Ministry of Finance, the Commission for Development and Reform, and the Ministry for the Development of Urban and Rural Housing issued a joint communiqué calling for the nationwide implementation of a public rental housing program following Chongqing's model. Then, in 2014, its attempt at relaxing *hukou* restriction was endorsed by China's first national urbanization plan, China's New-type Urbanization Plan, 2016–2020. This plan particularly encourages removing restrictions based on status and facilitating conversions of *hukou*, although this emphasis is only applied to medium- and small-sized cities, while controls on migration to large cities are reinforced.

Despite the central government's endorsement of Chongqing's approach, there is ample evidence at odds with the municipal government's claim and place frame of successful urbanization reform. While rural migrant workers are eligible for public rental housing, the randomness of the lottery system used for assigning units means that those who are most in need may not benefit from it immediately.

some cases. Rural migrants especially sought to escape poverty in rural areas and find opportunities in urban areas, which, combined with the need for laborers to fuel urban development, led to a sharp rise in rural–urban migration in the 1980s and 1990s (Chan and Zhang 1999; Fan 2008). Rural migrant workers have been key contributors to the urban economy: the majority of construction workers, service workers, and factory workers in Chinese cities are rural migrants (Swider 2015; Johnson 2017).

Although internal migration is now possible, having been nominally formalized in 1985 when cities started requiring migrants to register for temporary residence permits, the *hukou* system is still in place to control access to public welfare and urban services (Chan and Zhang 1999; Solinger 1999; Chan and Buckingham 2008). City governments now mostly control and define how the system works locally and how one might be able to convert their *hukou* (Chan and Buckingham 2008; Zhang and Tao 2012). This "localization of *hukou* management" (Chan and Buckingham 2008: 594), however, has offered limited benefits to rural migrant workers. It has mostly become easier for holders of temporary residence permits to acquire local *hukou* in small and medium-sized cities. Large cities, which are the main destinations for rural migrant workers, have implemented points systems that prioritize criteria such as education credentials, investment, and employers' qualifications to decide who can be eligible for converting *hukou*, creating high barriers for rural migrant workers to acquire local *hukou* (Zhang 2012; Zhang and Tao 2012).

An influx of rural migrant workers who predominantly work in low-pay jobs and the restriction on access to social programs and urban services based on *hukou* status lead to acute social stratification in post-reform Chinese cities (Zhang 2001; Sun 2014; Swider 2015). While rural migrant workers are literally the builders of Chinese cities, they are not seen as worthy of receiving benefits and services in cities (Zhang and Tao 2012; Swider 2015; Johnson 2017). Moving to cities provides few chances for social mobility, and the

The lottery system also breaks up communities and support networks that have already been formed among migrant workers. Similarly, the rural land exchange program, although enabling a set of rural households to financially benefit from restoring farming on their homestead, breaks up existing social ties as households in the same community make different decisions on whether to continue farming on restored land or use the payment they received to relocate to the city. The *hukou* reform is lauded for removing the two-tier citizenship in the city, but converting to urban status also requires rural migrant workers to give up their claims to the land they left behind in rural areas. For some rural migrant workers, this land is a form of backup plan, which they are not ready to relinquish. And this group has to continue enduring being excluded from the rights associated with urban status (except public rental housing). Likewise, not all rural households can or want to participate in the rural land exchange program, which arguably can lead to larger economic divides in rural areas. Furthermore, for rural residents living near the urban edge and at locations that are optimal for urban development, they have no choice but to face displacement and attempt to negotiate with the municipal government for a higher compensation. While some residents enjoy the benefits of the reforms and agree that they have helped address inequalities in the city, others see more drawbacks in or are excluded from participating in the programs.

The emphasis in this place frame on resolving previous challenges in urbanization, however, masks the shortcomings of Chongqing's reform. It also has additional political significance as it is closely attached to the arrival of Bo Xilai in 2007 as the party secretary of Chongqing along with his "Chongqing Model." Both of the programs discussed above were parts of Bo's "Chongqing Model," offering innovations in urban development for other Chinese cities that would be important for Bo's potential political success and career progression. Equally, framing the "Chongqing Model" as a more egalitarian mode of urban development also helped Bo gain popular support in the city (Zhang 2020). However, hereditary nature of *hukou* makes intergenerational social mobility difficult too (Chan 2010; Chan and Wei 2019). The lack of access to public schools also means many children of rural migrant workers are left behind in the countryside, creating childhood traumas from family separation (Ye and Pan 2011). The fact that low-income rural migrant workers tend to work in relatively unstable and temporary jobs, combined with their inability to acquire local *hukou*, pushes them into circular migration where they periodically move between destination cities and their hometowns or between different cities, making it more difficult for them to integrate at their destination cities (Zhang 2001; Fan 2008; Johnson 2017).

Scholars have pointed out that China's *hukou* system resembles international migration control (Chan and Zhang 1999; Johnson 2017). On the one hand, the entitlement to benefits and services attached to *hukou* makes the system function as a de facto citizenship policy (Solinger 1999; Johnson 2017). And on the other hand, the classification of "outsiders" as "deserving" and "undeserving" recipients of local *hukou*, especially in large cities, parallels similar international migration policies in several countries (Johnson 2017). In particular, the merit-based points system for converting *hukou* adopted by several large cities in China is very similar to points-based international immigration systems used by countries such as Australia, Canada, and the United Kingdom. The category of temporary residence permit holder that most rural migrant workers in China belong to also parallels similar "guest" workers programs used in Europe and the United States that bring in temporary and seasonal workers when needed, giving them the legality of temporary work and residence but no access to benefits, and send them home when their labor is no longer required (Roberts 1997; Johnson 2017).

See also related box "Freedom and Diversity in Cities" in chapter 3.

the arrest of Bo in 2012 on corruption charges and for the murder of a British businessman casts more doubt on the persistence of this place frame. For example, although the public rental housing program has continued since Bo's arrest, the housing is increasingly constructed in remote locations with poor infrastructure or is directly leased to factories, functioning as "state-subsidized workers' dormitories" (Roast 2019).

Conclusion

This chapter highlights how place framing can result from cities constantly being seen, understood, or positioned in relation to or in comparison with other cities. Different actors may construct different place frames of a city based on which context they situate the city in, what kind of position the city occupies in such a context, how they see the city relating to other cities, and so on. Viewing and understanding cities relationally means seeking out these different meanings, identifying those who help to produce them, and how and why these happen. The first Chongqing place frame explored here emerges from many individual visitors' social media posts about Chongqing, as well as by media coverage of the city. While it is unclear why Chongqing's mundane residential buildings and public transit infrastructure suddenly became the attention of tourists around 2016, the ability to post photos online and generate interest in these sites highlights and creates a discourse around their perceived uniqueness in comparison with most other cities. The subsequent international attention to Chongqing further enhances the framing of the city as a futuristic and cyberpunk one, shaped by an existing dynamic where large modern Asian cities are perceived as representing "the future" in western discourse. In both instances, this framing of Chongqing is an outcome of existing discourses about futurism and Asian cities and of ordinary people comparing the city with other cities, with average cities built on flatter terrain, and with both western cities and other similar Asian cities.

Land Markets

Land ownership significantly affects the urban built environment and urban development, and therefore is an important factor affecting urban planning and policy making. Its significance is also strengthened by the fact that land has long been treated as a commodity and an asset for generating revenue, raising financial capital, and speculation, playing a crucial part in capital circulation and accumulation (Harvey 1978; Massey and Catalano 1978; Christophers 2016; Ward and Swyngedouw 2018). China's remarkable urban development since the 1980s, especially the relentless transformation of its urban landscape, cannot be separated from the state's role as the landowner of urban land (Wu, Xu, and Yeh 2006; Hsing 2010).

As part of China's economic reform, land reform was initiated in 1988, with an amendment made in the Chinese constitution to separate land-use rights from land ownership. In this arrangement, urban land is owned by the state, whereas rural land is collectively owned by villages, and the use rights of land can be leased to users and developers, which are transacted in a land market (Lin and Ho 2005). This landowner role gives the state (specifically city governments) an important leverage to attract investment and shape urban development (Wu 1999; Ding 2004; Zhu 2005). An "industrial linkage and spillover" (Su and Tao 2017: 241) model is identified, where city governments lease land for industrial development at a lower price to attract investment, the spillover effects of which generate demand for commercial and residential land. City governments can then earn higher income from transactions of these types of land through competitive bidding (Xu, Yeh, and Wu 2009; Su and Tao 2017).

Land acquired an even more important role in China's urban development after 1994 when a tax reform re-centralized fiscal power that was decentralized to local governments in earlier reform (Liu and Lin 1998). As a result, local government income was reduced. As one of the exceptions, revenues generated from land transactions are still

Thus a city may be perceived by different actors through its relationship with or in comparison with other cities. Such positioning and comparison then generate particular framing or images of the city, which can have further implications for how the city negotiates with and is affected by the relationships and dynamics of its situation. In the example of the first place frame, the municipal government of Chongqing, while not having control over how the city's built environment is portrayed by outsiders, has decided to build upon and even contribute to enhancing the place frame. This decision comes from recognition that the place frame can make Chongqing appear exciting and appealing to potential investors, which has become a necessity for the city as it is situated in a global political economic network where different cities compete with each other to attract capital.

The second place frame discussed in this chapter, while coming from the municipal government of Chongqing, is also shaped by dynamics beyond Chongqing that in turn shape how Chongqing is positioned in relation to other cities in China. On the one hand, the increasingly heightened issues of urban–rural dual structure and inequalities in Chinese cities raised the need for finding solutions. And, on the other hand, Chongqing's special position in China's policy landscape as a strategically important city of the western region makes it possible for the city to experiment with policy changes. These two factors combined with one ambitious leader (Bo Xilai) led to the large-scale *hukou* reform, the public rental housing program, and the rural land exchange program, which provide the basis for the place frame of Chongqing as a city offering a more egalitarian mode of urban development. This framing is possible partly because of the existing national context and Chongqing's position in this context. It is also shaped by the municipal government's intention to change how Chongqing is positioned in relation to other cities: from being a leader of the western region to being a model for all Chinese cities. And this place frame has to some extent achieved this effect, given that two national policy changes since 2010 have more or less been modeled on Chongqing's completely under local government control, which makes land an important source of income for city governments, especially for funding the construction and maintenance of urban infrastructure (Lin and Zhang 2017). This change has been seen as a key factor that drives urban expansion, as local governments are motivated to use state power to forcibly expropriate land from rural collectives with low compensations to rural residents and to convert such land to urban land and a revenue source (Tao et al. 2010). Scholars have generally used the term "land-based finance" to characterize the heavy reliance on land for fiscal revenue by local governments in China (Wu 2021). A similar phenomenon can be found in the United Kingdom after the global financial crisis in 2008, where austerity measures cut local authorities' funding, making some of them choose to create arm's-length housing companies and invest in commercial property for additional income streams (Christophers 2019).

Increasingly, land is also used as collateral to obtain bank loans for funding infrastructure and development projects in Chinese cities (Wu 2021). Research identifies that city governments create arm's-length state-owned urban investment and development companies to circumvent the rule that bars city governments from directly borrowing funds on the market (Jiang and Waley 2020). These companies then use the land injected by city governments as assets, with the value of the land backed by city governments, to obtain loans to carry out development tasks, with the expectation that the loans will be covered by revenues generated from land transactions (Wu 2021). However, as these urban investment and development companies are state-owned commercial entities, they operate in opaque ways and raise concerns about how many debts they have accumulated that are guaranteed by and expected to be paid off by local governments (Pan et al. 2017; Jiang and Waley 2020).

See also related boxes "Urban Economic Processes" in chapter 2; "The Growth Coalition" in chapter 3; "Municipal Development and Finance Strategies" in chapter 4; and "Sprawl, Density, and Urban Growth Boundaries" in chapter 5.

approaches, despite the downfall of the leader associated with those policies.

In summary, the two place frames discussed in this chapter especially highlight how cities always exist and function in a world of cities with which they constantly interact (directly or indirectly). The relations with other cities, the positionality of a city in relation to other cities, and the dynamics underlying these relations all affect how a city is viewed by different actors, as well as how the actors associated with the city may act.

This chapter might help you to think about cities like Kolkata, India; Dubai, United Arab Emirates; Pittsburgh, PA, United States; and Valparaiso, Chile.

–7–
Jerusalem: Religious Tourist Destination/Ethno-National Citadel

Jerusalem is an urban place that embodies many contradictions in meanings, experiences, built environments, and physical landscapes. Some parts of the city are ancient, worn stone walls and steps. Others are modern and gleaming, steel and glass.

Simply locating Jerusalem by name is contested. "Jerusalem, Israel," immediately invokes counter(-national) narratives, because Jewish citizens of Israel as well as both Muslim and Christian Palestinians all claim it as their capital. Thus some observers will say that Jerusalem is in Palestine rather than Israel: naming either location is an act of place framing. Since its founding in 1948, the modern state of Israel has claimed Jerusalem as its capital, although at that time it only controlled the western side (while Jordan controlled the east side of Jerusalem, including the Old City and all of the territory of the West Bank, west of the Jordan River and the Dead Sea). In 1967 after the Six-Day War, Israel gained control of all of Jerusalem. Its residents include Jews, Christians, and Muslims, and its edges invoke stark juxtapositions between people living fully within the civil sphere and control of Israel, and those in a liminal space of long-time refugee camps and towns on the occupied West Bank, the Palestinian-administered (yet Israeli-controlled) side of the city's boundaries.

The difficulties of naming in Jerusalem extend not simply to the place and its geographical parts, but also to people. People who live in Jerusalem and throughout Israel and the Palestinian territories have a number of contested toponyms (place names), as well as demonyms (names for people) (see box, "Toponyms and Demonyms"). Israeli citizens can be a number of different religions (including Jewish, Christian, Muslim, and Druze), but their religions carry different citizenship rights and responsibilities to the state. Further, in Jerusalem, some residents who are not Jewish may have Jerusalem citizenship rights, but are not at the same time considered citizens of Israel. Naming these different groups of people is complicated if one wants to clarify that group's relation to the state at the same time; Israeli Arabs have Israeli citizenship, but may identify more as Palestinian than Israeli. But to call them Palestinian omits their citizenship status, and confuses them with people who are Arab/Palestinian but not Israeli.[1] (Consequently, we use terms in this chapter which seek to clarify both identity and citizenship: Jewish citizens of Israel, Palestinian citizens of Israel, and legal residents of Jerusalem.) The degree to which these competing national narratives are actively in evidence in the landscape depends in part on the awareness of the viewer about these different claims, and the sites/sights they see and acknowledge in the landscape.

Place conflict and radically different experiential narratives are unavoidable in this city. It may seem that Jerusalem stands – with perhaps only a few other cities in the world, such as Belfast, Beirut, Sarajevo in the 1990s, or Gibraltar – as a unique urban place because of its apparently irrevocable divides between its peoples and place frames. Jerusalem is less an exception, however, than a signifier of the constant particularities and specificities of any story, any theory, any place frame. Jerusalem and cities like it, with open conflicts over place meaning and boundaries, are simply the most explicit urban places, where the many competing strands of meaning and competing forces are explicit. Jerusalem also signals, insistently, the importance of power in crafting and maintaining urban

Toponyms and Demonyms

A branch of geography examines place naming, or toponymy, as symbolic of the identity and values of the namers (Zelinsky 1997, 2002; Rose-Redwood, Alderman, and Azaryahu 2010). While generally associated with cultural and regional geography, the issue of naming also gained prominence in US urban geography around the politics of renaming streets for civil rights icon Martin Luther King, Jr (Alderman 2000, 2003). A growing line of scholarship explores the power dynamics and symbolism of place naming, seeing toponyms as politically contested (Berg and Kearns 1996; Rose-Redwood, Alderman, and Azaryahu 2010; Rose-Redwood 2011; Alderman and Inwood 2013). As Alderman and Inwood (2013: 212) point out, toponyms are fundamentally about the "rights of people to participate in the production of place and to have their cultural identities and histories recognized publicly" (see also Rose-Redwood 2011).

Place names signify dominant identities within a region or city, and sometimes can be clues to struggles between groups over the power to name places. These struggles are only evident in active contestations, however, either moments when renamings occur or in historical records – some of which may be visible in the landscape, such as road signs with multiple toponyms. Attention to toponyms, then, means attending to who and what is dominant in a given landscape, and who/what is not. In Israel, toponyms have long reflected the prevailing power structure; in 1948, the modern state of Israel set out to "Hebraicize" (orient and render in the Hebrew language) the landscape, a particular linguistic nation-building exercise of toponymy (Azaryahu and Golan 2001).

Toponymy and demonymy – the naming of people – are both complex, particularly in Israel/Palestine, as demonstrated by the slash between those two place names, both applying to and claimed for the same territory. Using particular names for different groups of people in the country of Israel entails complex political decisions. Specifically, nomenclature for discursive meaning, with on-the-ground consequences for people's daily lives.

Because Jerusalem is contested and laden with uneven power and life chances, it is a useful site for thinking about the complexity and diversity of relational urban places: the constant juxtapositions, the hybridities and multiple scenes, the unexpected connections and sometimes irreconcilable narratives and place frames. Scholar Oren Yiftachel (2016) has argued that Jerusalem exemplifies the intersecting forces shaping urban places, pointing to structures such as colonialism, nationalism, religion, globalization, and gender as all having a role in producing the physical, economic, political, and social life in the city. In his rejection of a single urban theory to explain Jerusalem or any urban place, Yiftachel emphasizes structural complexity (what he calls "dynamic structuralism" [Yiftachel 2016: 485]) and temporal contingency.

In this chapter, we extend this attention to relational forces by examining the Old City of Jerusalem as an urban place that highlights the simultaneity of co-present but conflicting narratives and experiences. Jerusalem's Old City is just about a mile square, contained within stone walls, framing four distinct, named districts: the Armenian, Christian, Jewish, and Muslim Quarters each "belong" to, or exemplify, particular religious/ethnic/national identities. Although the districts are distinct, within the Old City the boundaries between them, while marked with signage, are nonetheless fluid. The identities of the Old City's districts reflect power shifts over millennia, and so Jerusalem also highlights time as an important element in the emergence and ongoing salience of place narratives. Jerusalem in particular signals the impossibilities of single narratives about places, the sedimentation of conflicting narratives, and the importance of consciously shifting frames to better understand urban places from different perspectives. In this chapter, we explore Jerusalem's *Old City as an inclusive tourist destination* of significance for three religions. We then contrast this celebratory framing by examining the underlying power dynamic of *Jerusalem as the capital of a specifically Jewish state*, erasing other claims and identities in the process.

> people who have long lived in the region and claim it as Palestine includes the terms "Arabs" and "Palestinian." Yet to call citizens of Israel "Palestinian" is somewhat oxymoronic, given that Israel/Palestine is a contested opposition between two competing nations, and doing so erases their Israeli citizenship status. Although Israel is explicitly a state for Jewish people, not all of its citizens are Jewish; they are also Muslim, Christian, and Druze. Israeli citizens who are Palestinian in ethnicity and nation-state orientation exist in a type of liminal impossibility of contradictory identities (Amara and Schnell 2004). Furthermore, Palestinians live in many different states, none of which represent a "home"; some are citizens of Israel, others of Jordan or Lebanon, and millions of others live in Gaza, a territory without state identity, or the West Bank, territory controlled and occupied by Israel and without formal state status.
>
> In Jerusalem, it is particularly important that demonyms help to clarify the different statuses of Palestinians – especially whether they are considered legal residents of Jerusalem, citizens of Israel, or non-state Palestinian residents of the Israeli-occupied West Bank. (All of these specific groups of Palestinians live within the Jerusalem area.) It is generally recognized, but not universally agreed, that Arab citizens of Israel prefer to be called "Arab Israelis" (Amara and Schnell 2004; Schnell and Haj-Yahya 2014), which clarifies their citizenship status but does not recognize a national identity separate from Israel.
>
> *See also related box "City Branding" in chapter 6.*

We explore Jerusalem's Old City through a touristic lens in order to constitute the place, as visitors to any place often do, by bundling images, sensory, and cognitive experiences into a place assemblage. In doing so, we focus on the processes by which one comes to a place with particular lenses that "see" some aspects of the city more than – or to the exclusion of – other aspects and narratives. While tourism may seem to be an overly surficial perspective on urban places, and this place in particular, it highlights the partiality of knowing, the constant filtering of facts and sensory experiences that produce incomplete but meaningful place knowing. Tourists arguably know that their place experiences are partial, recognizing that different experiences, stories, and landscape elements converge with what they see and experience to constitute any given place more completely. A touristic approach also echoes the scholarly approach of extracting empirics and processes from the complexity of urban life in order to gain some distilled understanding of places.

Tourists, wherever they may be, represent privilege, the ability to explore with leisure, the freedom from regular day-to-day labor and obligations. Tourism or idle people-watching reflects a consuming gaze within and of a place, and is central to a certain understanding of knowing cities. Being a tourist, or flaneur, entails choice – a choice to travel, to watch, and to engage in discovery, and a choice of location. The decision to visit Jerusalem's Old City signals a positionality of privilege and comfort with *this* place at *this* time. It is largely to accept – for some, even to embrace – the status of Jerusalem as the capital of Israel, fully controlled by the Israeli state, illustrative of the identity and realized aspirations of Jewish people for a nation-state. Many people who have the choice to be tourists may not choose this place. Some would-be tourists may have concerns about personal safety. Others object to the unresolved status and oppression of Palestinians living under Israeli military occupation in the West Bank – including parts of Greater Jerusalem. Some people may wish to visit Jerusalem but not be able to obtain permission to enter Israel, due to Palestinian identity or political activities that the Israeli state deems hostile to its existence.

Touring the Old City

Most tourists come to Jerusalem via the notoriously busy route to the city's west (from the direction of Israel's international airport, and Tel Aviv–Jaffa city on the Mediterranean coast). The four-lane highway climbs steeply in elevation as it nears the city, and one attentive to the landscape can see buildings perched atop hills at shocking angles to the road. The landscape gets drier, and the roadside more brown, with the elevation. Along some parts of the road are signs which mark a particular history, noting places where the *Haganah*, or Jewish paramilitary organization in British Mandate Palestine, fought over a blockade of the city in 1948, noting breached sites. A particular Israeli nationalist narrative makes claim here, linking travel to the city to the history of the modern state and its continuous assertion of the centrality of Jerusalem to the state itself.

Once off the highway and onto city streets, the city, like any other, offers a cornucopia of sights, with all manner of dress and styles. From the windows of a car or tour bus, seeing people walking or standing on sidewalks, the juxtapositions of multiple forms of life are visually arrayed. Fashions vary from the long beards and black coats of orthodox Jewish men to secular women in high heels and short skirts; a wide variety of young and old people sit at bus or light-rail stops, holding leather bags, colorful plastic reusable totes, and store-branded flimsy plastic bags. The visible place frame here seems to be diversity and coexistence, at least among different types of Jews.

The main road in the direction of the Old City parallels the modern light-rail line. Buildings are mostly low rise, with sandstone exteriors but also with plenty of glass and steel in evidence. The long history of this place is visible in its architectural variety. Sidewalks, particularly on side roads, are marked with low bushes and some trees.

The Old City of Jerusalem is the historic core of the city, surrounded by stone walls. It draws its cultural power and global attraction as the site of key landmarks for multiple world religions: the Western Wall of the old Jewish Temple; the Islamic Mosque Dome of the Rock, from which the spirit of Mohammed was said to ascend to heaven; and the Church of the Holy Sepulcher, sacred to Christians as the site of Jesus' tomb. The very fact of these multiple, religiously contrasting sites in about 1 sq. km of the world's space highlights the multiplicity of meanings, the overlapping *places* of this location, and its significance for people all over the world. Jerusalem's Old City, for tourists to these sites, may seem like *an inclusive tourist destination*, one which is always fragmented by one's perspective, position, and willingness to think outside a given framework.

In present-day Jerusalem, tourists in particular tend to be viewing the city through one of two lenses; that of Christianity or that of Judaism (both containing multiple strands reflecting different sects and practices). Getting to the Old City as a tourist usually means navigating the crowded, frenetic city streets, from where Route 1 ceases to be a highway, to a stopping point close to an Old City gate. In West Jerusalem, visitors might ride a bus or drive up close to New Gate or Jaffa Gate, both surrounded by stone or concrete buildings close to the sidewalk, hotels, shops, and hospitals. Or they will have walked from a hotel in the vicinity of the Old City. A common entry point is Mamilla Mall, a pedestrian shopping concourse (with a parking garage below it) that opened in the first decade of the twenty-first century, tucked into redeveloped masonry buildings from the nineteenth century, replete with tall arches. Nestled in the arches are the large windows and distinct signage of many globally recognizable high-end retail stores and some Israeli chain stores.

Mamilla presents visitors on their way to the Old City with an internationally familiar consumer scene as a gateway to an ancient, sacred landscape, yet one that also teems with consumption opportunities. Mamilla Mall also offers juxtaposition of old and new in both its physical manifestation, with large plate glass, colonial arches, stone masonry, and in the mix of people walking, sitting, and using the space. Orthodox Jewish families are apparent in their modest dress, pushing strollers, male children with telltale fringes/tassels of *tzitzit* (traditional underclothing) hanging beneath their outer clothing; groups of women stand out as nuns with their habits and long skirts; while other groups wear thoroughly contemporary dress. Palestinian citizens of Israel are often among this latter group, frequenting the mall both to shop and as workers (Shtern 2016). At the east end of the Mall, a large staircase (and elevator off to the side) enables pedestrians to climb the hill to the Jaffa Gate entrance of the Old City itself.

Up the stairs, the imposing sandstone masonry walls of the Old City loom beyond a wide concourse leading to the Gate, forming a pedestrian bridge over a busy street below. The walls of the Old City are open here, allowing access to cars and small trucks. Pedestrians can enter adjacent to a tower, walking under an archway. Much of this area is marked by movement – cars below the bridge rushing past the old city, people flowing in and out of the gates. There are also moments of stillness. Tourists stop to take photos; some people linger (perhaps to meet someone or just take in the sights), and soldiers with rifles watch the flow at the Gate itself, eyeing people as they go by or chatting to each other. Sometimes musicians or other entertainers stand along the concourse expanse; people stop to watch and listen.

Once inside the Gate, this space of the Old City feels open, with a courtyard-like space where people mill about, cars are parked, and some shops have tables and chairs out for people to sit for coffee. Postcard racks are lined up outside of many stores, and signs in English, Hebrew, Arabic, French, German, and other European languages indicate where tourists can buy tickets for access to the rampart walk on the city walls or a tunnel tour. Two roads lead in different directions as means to enter more deeply into the Old City itself, since the Jaffa Gate marks the border between the Christian (left side) and the Armenian (right side) Quarters of the Old City. Tourists may mark their own identities in part by whether they take the road to the right, into the Armenian Quarter and then heading east toward the Western Wall through the Jewish Quarter, or straight ahead, walking between the Christian and Armenian Quarters toward the Muslim Quarter. While people can explore and access all the Quarters regardless of their initial path, many tourists tend to orient themselves either toward the major Christian sites, such as the Church of the Holy Sepulcher where Jesus is said to be buried, or toward the Western (or "wailing") Wall, the only remnant of the retaining walls of the First and Second Jewish Temples in Jerusalem (the Second Temple having been destroyed in 70 CE by the Romans).

Regardless of direction, the paths at this point evoke a city of a more ancient time, the cobblestone roadways oriented largely to pedestrians (despite the occasional vehicle such as a delivery van making its way through the narrow streets). Shops are close together, their wares lying out on tables and hanging in the large open doorways. Overhead, awnings and perforated metal roofing – which keeps birds at bay – block out some of the sunshine, fostering a feeling of being partially indoors (although on rainy days, it doesn't keep the rain from filtering through). As a tourist, it can feel as though all of the stores are catering to a yen for trinkets and mementoes, with shops selling keychains, bags, hats, jewelry, scarfs, rugs, leather goods, and more. There are also shops that are clearly oriented to locals, with produce, groceries, and mounds of brightly colored ground spices like paprika, turmeric, and saffron. Depending on which corner one turns, the trinkets may have more Hebrew or more Arabic writing, more Christian, Muslim, or Jewish iconography, and foodstuffs vary as well. Turkish coffee and lamb kebabs along the Via Dolorosa combine with Christian imagery on a street known as the path that Jesus took to his crucifixion. Here many of the tourists make their way to the Church of the Holy Sepulcher to roam with the crowds in the site where Jesus is said to have been both crucified and resurrected. The church complex has a large plaza in front of it where tourists move around in seemingly organized groups, some clearly with guides. Dress here ranges from formal religious garb denoting nuns and priests – including the distinctive miter hats of the Eastern Orthodox – to casual shoulder- and thigh-exposing summer wear of many non-religious visitors and locals.

At certain times of year, the flow of pedestrians going to and from the Old City can seem more like a human crush. These include major Christian holidays such as Easter, when the Via Dolorosa and Church of the Holy Sepulcher are particularly packed, or one of the three Pilgrimage holidays in the Jewish calendar (Passover and Shavuot[2] in spring, and Sukkot[3] in the fall), when large numbers of Orthodox Jews dominate the groups of people going into the gates and heading toward the Western Wall to pray and put notes to God in the wall's crevices. No matter the season, Jerusalem's crowds offer visible mixes of Orthodox Jews, Christian groups of nuns and Eastern Orthodox priests, and the more numerous and seemingly secular and thoroughly modern tourists or other locals dressed according to the weather rather than religious mores.

At the Western Wall, or Kotel, the space has a distinctive nationalistic feel, contrasting the multiplicities of religions and identities that mix in the Old City (albeit each dominating a particular Quarter). People who have accessed the area by walking through the Old City emerge in a small area that looks down on the Wall and a large plaza in front of it, where the Israeli flag flies prominently and the space is uncharacteristically open compared to the rest of the Old City. To get to the plaza, one has to pass through security gates and show the security guards the contents of any bags being carried – the latter an experience common for anyone who has entered a public indoor space in Israel – and then walk down a set of steps to the plaza bottom.

Even in crowds, the larger space of the plaza in front of the Kotel gives the area a very open feel. At one time, it was as fully and densely built as other parts of the Old City and known as the Moroccan Quarter. The area was demolished – and its residents displaced – after the 1967 Six-Day War, at which time Israel extended its sovereignty over all of Jerusalem, including control over the Western Wall. The extra security, flags, openness, and prayer spaces directly facing the Kotel all contribute to marking the space – despite its proximity to the al-Aqsa mosque above and behind it – as specifically Jewish and Israeli. Indeed, the Israeli authority and ethno-nationality of this space, and of the broader city, are highlighted here, offering a stark contrast to the density and diversity of the rest of the Old City.

Once down some stairs to the plaza, there is room to walk around and to access the separate women's and much larger men's prayer spaces directly facing the Kotel. The gender-separated spaces offer a glimpse to a wider conflict within Israel, which is the authority of Orthodox rabbis in civil society. At the Kotel, as in Orthodox synagogues, men and women are separated by physical barriers, even though other traditions in Judaism recognize women rabbis and other prayer leaders and allow men and women to pray together (see image). A multi-denominational group of Jewish women protest this segregation regularly, seeking fulfillment of a government promise to provide space for women to pray as equals at the Wall. The presence and demands of this group of women highlight exclusions and power struggles within Israeli society.

Behind the Western Wall, the Dome of the Rock looms above, part of the Temple Mount complex with the al-Aqsa Mosque. The Western Wall is technically part of the retaining walls surrounding the Mount, where the first and second Jewish Temples were built before they were each destroyed. Although the Dome is visible behind, the focus of the plaza space is the Kotel, the Israeli flags, and Jewish religious prayer spaces in front of the Wall.

The Kotel offers one lens and narrative on the meaning and significance of Jerusalem's Old City. But not everyone, even tourists, is oriented to or interested in the Kotel. Entry to the Old City via a different gate offers a different perspective on who is coming into the area and why. The Damascus Gate on the north side of the Old City, for example, is adjacent to neighborhoods where Palestinians live (these residents may have Israeli citizenship, but not necessarily). The areas on both sides of this gate, beyond its plaza and busy street, are consequently more Palestinian in character, as evidenced by more Arabic writing on the signs of shops in the area, and the outer robes or headscarves worn especially by older Palestinian men and women. The Damascus Gate offers a wide, amphitheater-shaped entry to the Old City, with stairs in multiple directions from a busy street marking the opening to the Old City. This plaza functions as a gathering place especially for Palestinian residents of the city. The entrance to the Gate itself is a bridge entering into an arch in the Old

Jerusalem

Young boy at the Western Wall[4]

City walls. The bridge overlooks an excavated Roman-era gate in a lower section of the wall. Entry at this ancient lower gate reveals a small display of some of the excavated columns and spaces below, exposing another time and layer of Old Jerusalem. The Roman Gate underneath the present-day Damascus Gate actively signals the layered meanings and power structures of the city, which endure.

The Damascus Gate also signals deeper divisions and tensions, and the present-day uneven spatial authority and control in the city. In spring of 2021, for example, control over space and expressions of territorial dominance were particularly overt when Jerusalem police cracked down on public gatherings at the Damascus Gate during the Muslim holy month of Ramadan. It was not clear why traditional Ramadan evening gatherings were particularly problematic that year, but the holy month did coincide with two significant Israeli holidays, Memorial Day and Jerusalem Day, the latter commemorating the 1967 Six-Day War, when Israel took control of all of Jerusalem. The competing identities of Jewish and Muslim in Jerusalem, which most days uneasily coexist, became intractably in conflict in spring 2021, contributing to an eruption of violence – a recurrence of open war, even – when Jewish citizens of Israel who were celebrating Jerusalem Day clashed with Palestinian youth, and Hamas (in the Gaza Strip) bombed Jerusalem as well as other cities in Israel. Israel retaliated with an extensive bombing campaign against Gaza. Although a ceasefire was declared, the tensions that precipitated the violence remain.

The Damascus Gate and its literal underlying ancient gate highlight the millennia of conflict and changing power structures in and over Jerusalem (see box, "Landscapes and Power"). Contemporary visitors, especially tourists, to the Old City, by their presence, signal acceptance – indeed, often active endorsement – of the existing status quo in the city. They represent people who can and choose to travel freely within this space; they seek access to it either due to general interest, religious connection, or other personal connection. The sites they choose to visit highlight the viewpoints or histories that they wish to engage.

Landscapes and Power

A key concept for geographers and others who think about place and space is landscape, the material elements of a given site or scene, including both natural and human components. Geographer Carl Sauer articulated the concept of landscape in 1925 as a specific empirical object of the study of human and environmental relations (Sauer 1925). He called it the "cultural landscape" and argued that studying landscapes would help scholars understand how humans adapt to and change the environment. The concept was not initially articulated or applied to urban areas, but geographers and other scholars have used the concept to talk about how societal values and power are embedded in, and symbolized by, landscapes (Lewis 1979; Schein 1997; Mitchell 2003). Power in landscapes is symbolized by monuments and other forms of memorialization, as when societies confront or deny their own histories of genocide (Leib 2002; Till 2005; Dwyer and Alderman 2008). It is also symbolized in the everyday, by what is constructed and what isn't, and through behaviors considered acceptable, and those not, in certain places (Mitchell 1997; Blomley 2003).

Landscapes are constituted and maintained through social norms but especially through laws such as zoning and other forms of property or people regulation (Schein 1997; Mitchell 1997; Blomley 2003; Martin and Scherr 2005; Valverde 2012). Thus regulatory operations of power are hidden from explicit view in the resulting landscapes, normalizing or hiding the relations that produced particular built environments. For example, many American university campuses were originally built and maintained by enslaved people; rarely are their labors and lives acknowledged in any form in the resulting landscapes (Inwood and Martin 2008; Wilder 2014).

Landscape scholarship examines power differentials and transformations that have produced them: Martin and Scherr (2005) argue that rather than being fixed, landscapes are opportunities – or fields of action – for different actors

One of the "other" viewpoints explicitly marketed to visitors to Jerusalem is the partial "bird's eye" view from the top of the ramparts (the Old City walls), which exposes the palimpsest of the Old City through historical markers and glimpses of contemporary complexities. The ramparts tour allows people to walk, self-guided, along the top of the Old City's walls, offering a distinctive perspective on the spaces of the Old City. Whereas the city from its narrow and slightly dark, crowded streets seems completely jam-packed with people and buildings, from the ramparts one sees open spaces: small courtyards and gardens, a basketball court here, a soccer field and school playground there. Laundry is visible drying on lines hung between buildings and across windows. Satellite dishes abound, as do white cylindrical water tanks, some with solar panels visible next to them, in the jumble of uneven rooftops. Looking outward, away from the Old City, the wider roads with sidewalks, hills with trees, and taller buildings all contrast with the density of buildings and activities inside the walls. Informative signage offers details about the buildings visible from different points, and the history of the constructions of various gates and towers by different city leaders over a millennium.

By contrast, tourists may also go underground to visit the Western Wall tunnels. This tour focuses more directly on the architecture and history of the Jewish Second Temple, dating a few hundred years before the Common Era (CE) until its destruction in 70 CE. Visitors to the tunnels go below the section of the wall that survives above ground, seeing a far more extensive architecture that includes ritual baths, water trenches, and a quarry. This viewpoint on Jerusalem offers an archeologically grounded narrative of Jewish existence in this place for millennia, giving tourists a place frame of life in the past that simultaneously makes claims on the present.

The viewpoints from the tunnels and the ramparts emphasize the partialities of the tourists' experiences and knowledge of the city – any city. They also point to the ever-changing power structure in Jerusalem – at various points in time, the city was controlled by, among others,

> to shape both practice and the more material forms of social life. In this sense, transgressive or alternative uses of space, such as sleeping in public urban spaces (Mitchell 1997), use of temporary displays (Inwood and Martin 2008), and marching through public streets (as in Marston 2002; or Allen 2020), can not only highlight how some groups and forms are expression are marginalized or oppressed within existing landscapes but also push to transform these dynamics.
>
> The limits to landscape transformation are embedded in their material form; while space can be transformed to be more broadly inclusive, such changes require mutual accommodation. Not all spatial or landscape needs and expressions can coexist because of the real limitations of physical space (Pierce 2022); thus landscapes, as spaces and places, are sites of ongoing contestation and struggle.
>
> See also related boxes "Urban Cultural Geographies" in chapter 2; "Urban Environments" in chapter 4; "Black Geographies" in chapter 5; and "City Branding" in chapter 6.

Jews, Romans, Ottomans, British, Jordanians, and Israelis. Being in the tunnels highlights not just how old the Old City is, but how multilayered it is physically, and also socially and politically. It does not sit alone of course; Jerusalem's Old City is surrounded by the expanded built-up parts around it. These include the crowded and contested largely Palestinian neighborhoods arrayed in the valleys to the north, east, and south of it, and the expanding, modern, and Jewish-dominated western neighborhoods. These areas offer complex and contradictory identities and political statuses, most significantly depending on whether the neighborhoods are understood and controlled as part of Israel, or part of the contested, occupied, and differentially managed Palestinian West Bank. These distinctions and their social and political meanings are largely invisible and unremarked by most tourists, but they define the place frame of power and control which situates and enables the tourist experience.

Framing power in the city

Being in the Old City of Jerusalem as a tourist is, in some ways, to accept and suspend certain tensions about spatial and territorial power and authority. The Old City teems with multiplicities: peoples, religions, and languages. These are not merely the signifiers of a world tourist site; they manifest the real presence of multiplicity every day in the city, in its residents, its institutions, and businesses. Street signs in Hebrew, Arabic, and English portray the multi-layered history, with "Omar Ibn El-Khattab Square" just blocks away from "Greek Catholic Patriarchate Street" and "Maronite Convent Street." At the same time, along the walls of the Old City, on the patches of its security forces, and in open spaces, the Israeli national flag is prominently displayed. These latter symbols and presences highlight the importance of *Jerusalem as the capital of a specifically Jewish state.*

The dominant visual narrative, or place frame, in the Old City of Jerusalem today, in the 2020s, is one of relatively uncomplicated Israeli control. Yet in a day-to-day sense, this control is continuously complicated by sporadic conflict and violence, and by the ongoing presence of Palestinians, many of whom articulate nationalist claims to space within and adjacent to the Old City and Jerusalem as a whole. While any narrative or frame of a place represents the authority or dominance of one lens and position over another, this dominance – and the presence of counter-narratives – is particularly evident in Jerusalem.

To experience the Old City as a tourist, or even, indeed, as a Jewish citizen of Israel, is to absorb and partake of a narrative of Israeli triumphalism, grounded in seeing Israel as the appropriate national homeland for the Jewish people, with Jerusalem as its capital. In this viewpoint and place frame, Jerusalem – particularly its Old City – represents autonomy, belonging, and national pride. Yet for others, including some tourists but most significantly for Palestinians, both with Israeli citizenship and without, this manifestation of the Old City, with its Israeli flags and explicit security presence, signifies marginality and oppression.

Jerusalem is a city perhaps more than any other (although it is sometimes compared to Belfast in Northern Ireland) with place frames dramatically, painfully, and violently in conflict. In contrast to the Israeli national narrative about Jerusalem as the unabashed capital of Israeli and Jewish identity; another place frame of displacement, containment, and incompleteness opposes the first – along with counter-claims to it as a capital of a future Palestinian State. About 20 percent of Israel's overall population is Palestinian (usually called "Arab" within Israel) who are a mix of Christians and Muslims, approximately 350,000 of whom (about 18 percent of the total) live in the neighborhoods of East Jerusalem.[5]

Among the central conflicts in Jerusalem between Jewish and Palestinian identities is that over land claims. Both Jewish and Palestinian residents of the city were uprooted in the 1948 war waged after Israel declared independence. At the war's end, Israel maintained control over West Jerusalem whereas Jordan controlled the Old City and East Jerusalem neighborhoods. Many Palestinians fled to Jordanian territory because they were afraid of staying under Israeli control; or expected to be able to return after Israel was reconquered; or were forced out of Israeli-controlled areas by Israeli military forces (all three of these dynamics were at play in different areas in 1948). After Israel captured all of Jerusalem, the West Bank, Sinai, and Gaza after the 1967 Six-Day War, the government allowed residents of East Jerusalem neighborhoods who were Jewish and had been displaced in 1948 to reclaim their homes, while Palestinians were not allowed to make such land claims. Palestinians living in areas of Jerusalem formerly controlled by Jordan were granted permanent Israeli residency, but not full Israeli citizenship (unlike other Palestinians in other parts of Israel, who are Israeli citizens); this residency is contingent on continuous "physical presence" in Jerusalem.

The spring 2021 war between Israel and Hamas hinged in part on the active contestation over land in the East Jerusalem neighborhood of Sheikh Jarrah, where Palestinian families who had lived in western Jerusalem before 1948 were resettled by Jordan between 1948 and 1967 (Krauss 2021). Jews who had owned that land, and were displaced in 1948, sought to reclaim it after 1967. These land claims have been contested for years and, along with Jewish purchases of land and settlements in East Jerusalem neighborhoods, form a flashpoint of contestations and at times violence within the city and sometimes beyond it.

In the Jerusalem area, both within the city boundary itself and just outside of it, Palestinians live with up to three distinct legal statuses (Mitnick 2015).[6] One group, Palestinian citizens of Israel, have Israeli national citizenship. The second, permanent Jerusalem residents, are Palestinians from East Jerusalem neighborhoods (or the Old City) who are legal Jerusalem residents but not Israeli national citizens. Finally, Palestinians who are non-Israeli citizens live in neighborhoods of Jerusalem that are in the greater West Bank area, under the administration of the Palestinian Authority. Jerusalem residents with permanent resident status are able to live and work within the city limits of Jerusalem and can vote in municipal elections (but not in Israeli national elections, unlike other Palestinian citizens of Israel who live in other parts of Israel). Palestinians who are not citizens of Israel and who do not have permanent Jerusalem residency are not allowed to enter Jerusalem without a work permit or other special (such as medical) permission, despite the fact that in many cases their towns are immediately adjacent to municipal Jerusalem. Distinctions between these groups of Palestinian residents of Jerusalem are evident in the identity cards that they carry (blue for Jerusalem city residents, green for West Bank residents) and the life opportunities accessible to them. The demarcations of these two groups are evident in the landscape via the separation wall that snakes through the hills and valleys around Jerusalem's eastern, northern, and southern boundaries, and checkpoints which regulate crossing from one side to the other.

The physical, social, and political separations and distinctions within Jerusalem are not immediately or even particularly evident to most visitors or tourists. (Indeed, the largely Jewish residents of mostly West, but increasingly also East, Jerusalem neighborhoods likely do not notice these separations to much degree either because the separations do not affect their daily lives.) The impression and feel of the Old City, in particular, is one of cross-religious and cultural mixing and mobility. Yet there are glimpses, such as in a view from the Old City ramparts, of the separation walls that line some of the hills and valleys, and which enforce the broader segregation and power structures of this place. From this viewpoint, however, the wall is too far and small a view to see how such a line in the landscape shapes lives and livelihoods. Only a visitor determined to understand the divisions of the city, and lives of Palestinians, would venture into specifically East Jerusalem neighborhoods, and few would cross into West Bank neighborhoods on the other side of the wall. If they did so, they would see neighborhoods characterized by older, in some cases evidently run-down, overcrowded buildings. Permits for new construction are extremely hard (functionally impossible) to obtain for Palestinians within Jerusalem, and so buildings are old and in decline and support too many residents. Walls within these areas are replete with graffiti about Palestinian rights.

Conclusion

The Old City of Jerusalem sits within these broader realities as an emblem of a *place made for* certain kinds of view and experiences. It is a place made for visitors to marvel at the juxtapositions and seeming compatible coexistence of three world religions. A place for local residents to – unevenly – daily claim their places, their beliefs, their belonging. It is a place made for a national set of claims, impositions, and celebrations.

Jerusalem contains many deeply contradictory meanings and landscapes, the central dynamic of which lies between its status as capital for one nation – Israel – while simultaneously claimed as capital to a second people – Palestinians – who seek nation status and live in insecure and fragmented territories. For the city to exist and even thrive as a tourist site, as it has for the past more than 50 years with relative stability, has required many layers of separation, surveillance, and imposed order. One aspect of that order has been to maintain the Old City as a sacred and also accessible site for three different religions, each with its own internal subdivisions.

The Old City of Jerusalem offers a stark and extreme reminder that knowing and understanding cities means recognizing that conflicts are embedded within them. Sometimes those conflicts are better understood as historical layers; lives lived and then replaced, sometimes by a bulldozer in a violent reworking, and sometimes by gradual changes, erasures, and new constructions. In Jerusalem's Old City, these layers are less palimpsest – faded pasts overlain with a dynamic present – and more active, bitter, and irresolute place framing – the presence of one reality that denies and obstructs another. The spatial incompatibilities inherent in Jerusalem do not usually manifest in an overt, daily impossibility; rather, they sit uncomfortably, enacted by walls and security forces, historical plaques denoting settled narrative, and the ongoing persistence of daily life practices of mobility and exchange (economic and social). It is in the practices of daily life – whether that be in the mundane of provisioning or the sacred of prayer, or the push of alternative narratives through gathering and protests – that the city as a place is made and remade. Understanding these moments of place-making means actively seeking out the overt and covert framings that give life and built environments their narratives, if not coherence.

The Old City of Jerusalem offers a lesson about urban knowledge; it is always fragmentary and incomplete. What is visible and evident – stone walls and streets, people engaging in commerce and interaction adjacent to sacred sites – exposes place frames which not only juxtapose each other but can undermine and contest other framings, or understandings, of place. In order to begin to make sense of cities, we have to ask questions about who has power, what other stories of this place exist, and how these stories interact and why.

This tour of the Old City of Jerusalem illustrates and highlights that many of our surface observations of an urban place are ones that reflect dominant place frames. To fully, or at least better, understand a city is to keep trying to look beyond the obvious impressions for other meanings or place frames.

This chapter might help you to think about cities like Mecca, Saudi Arabia; Beirut, Lebanon; Belfast, United Kingdom; and Nicosia, Cyprus.

8
Conclusion: The Impossibilities of Fully Knowing a City

The past few decades have been momentous for urbanism globally. In 2007, the United Nations estimated for the first time that more than half of all humans lived in cities. At the time this book went to press, about 15 years later, there were roughly one billion more urban residents than rural ones, and the margin is growing rapidly.[1] In the United States and Europe, more than three-quarters of all people live in urban areas. Urbanism now shapes the majority of forms of life for people worldwide.

Yet what "the urban'" means varies substantially across the globe. Around the world, a majority of the new growth in urban population each year lives in slums of the Global South. In other words, the characteristics of northern urbanism – intensified infrastructure, services, cosmopolitanism, wealth – are not recognizable as normal in most new urban growth. Whatever it means to think about cities, the project of urban analysis has to be useful in contexts where both the lived reality and the aspirational ideals of urban residents vary wildly from the experience of any particular reader of this book. We need a framework that allows us to admit many different ways of "being urban" toward many different ends.

Thinking about cities rigorously means trying to consider lots of kinds of things at once. These include the physical form of the urban landscape, its aesthetics, the types of jobs and companies located within it, the histories of development that favor some people and spaces over others, the lakes and trees and parks where people walk, run, and gather, and the networks that connect people in one city to other people and places. In this book, we have argued that it is important to think about cities as contexts for place-making, where people negotiate over the meaning of the city, the shape of their lived geographies, and the future of the built landscape. Urbanites argue about which things are part of particular places; in doing this, they define them. These arguments constitute place framing, and they have real consequences for cities and the world.

As we noted in the Introduction, cities are sites of both propinquity and density. Propinquity means that cities are sites conducive to productive interactions; people are prone to make useful connections even when they can't see the network of relations from which those connections arise. Density is evaluated relative to a context: generally, cities are more densely settled and built up than the areas around them or between them. Yet saying that cities are dense and propinquitous does not mean that we understand urban places' evolving logics, which can vary widely. To know what matters about a city, one has to pay attention to what people argue about when they're trying to make it.

In each of the case study chapters of this book, we have posed at least two ways that stakeholders see the cities we've discussed: frames for which different constituencies argue as they try to make a city reflect their distinct visions and desires. In some cases, the place frames we present are not in sharp contradiction but instead are merely orthogonal to each other. They offer different visions, but one doesn't confront or undermine the alternative frame discussed. In other cases, though, adopting one frame does subvert another.

Competing place frames reflect distinct perspectives and positions on the processes that produce cities. This can include political imperatives at differing scales (as in Worcester), or economic wealth concentration that accompanies or even produces certain cultural and lifestyle choices (as in the City of London or Tehran). In each of the cases we've discussed, though, there are many more place frames than we have narrated. Indeed, the people who live and work in cities place bundle them in a wide array of ways, only some of which connect to or cohere as shared place frames that can be seen in public discourses and practices.

Our point in selecting various frames isn't to say simply that one can know cities in different ways. It is that we cannot hope to explain these cities' complexity and diversity by centering on one single overarching urban theory or trope. Furthermore, we cannot "add up" all of the frames we mention to get a "whole" place. Instead, while it is true that a place-framing approach helps to see the many different dimensions of a place, it also emphasizes how these different dimensions are always being claimed or rejected by various parties. Part of the point of a place frames' approach is to be able to excavate the many partialities of urban places *without* claiming that, as a result, a city is fully known.

Place framing is an essential part of urban politics. Place framers' actions reproduce patterns of structural control, while other place framers struggle to resist that control and assert other social and geographical patterns in a city. The most visible, most highlighted, most shared frames in a specific urban context inevitably occlude or disrupt the consolidation of other place frames. Place politics is an always ongoing process of *emplacement* (the solidification or consolidation of place frames by relationally empowered actors) and *displacement* (the disruption or erasure of place frames for which relationally less- or dis-empowered actors struggle and fail to win consensus or even visibility).

The importance of interrogating cities

The authors of this book have been writing for most of two decades about place framing and relational place-making. We argue in our scholarly work that a place-framing approach helps to usefully interrogate the diversity of urban forms and urban lives that we see in actually existing cities. Both separately and together, we've written about urban place-making in North America, including: St Paul, MN; Athens, GA; Worcester, MA; Boston, MA; Pittsburgh, PA; Atlanta, GA, Toronto, ON; and Montreal, PQ. Fundamentally, this work has always been about politics; we have sought to understand and explain urban conflict as disagreement about how cities should look and the modes of life they should enable. Often, our writing has centered on specific instances of activism, such as neighborhood organizations trying to improve quality of life on particular urban blocks (Martin 2003b) or homeowners and renters trying to stop land-use change in their neighborhoods (Martin 2003a).

Writing about place has often been understood as particularist: places are characterized (often positively) as local in contrast to the global (Cresswell 2019), and analyses of urban places are sometimes characterized (critically) as parochial rather than generalizable (Harvey 1996). Yet we believe that our view of place – as relational and political – is *crucial* to understanding how cities work. We have not always said so as clearly as we do in this book.

While our past work has often been in cities, it has not always been explicitly about urban processes. Yet we have long believed that it is useful to notice how urban processes are plural and that perspectives on the city are always fragmented. We have been drawn to studies of place politics precisely because we believe that these politics are *the best* way to understand the diversity of processes that play out in cities. Place politics reveal the values of the people who actually live in, and use, cities, as well as those who avoid and undermine them from afar. Place framing isn't just useful for understanding specific instances of place/land use politics. We understand the urban condition as an always ongoing process of place-making through place framing (Martin 2003b).

Many urban scholars use Henri Lefebvre (1991 [1974]) when they try to think about the plural dimensions of the urban condition. We agree with Lefebvre that the terrain or geography of cities is best understood as part of an ongoing process of production. Lefebvre sees urban space as a multi-dimensional triad: it is simultaneously perceived, conceived, and lived. Each of these three dimensions signals different aspects of individual and collective processes. He is very clear that these aren't separate phenomena in practice but facets of the same urban space.

We've written about the important, partial parallels between Lefebvre and Massey before (Martin and Miller 2003; Pierce and Martin 2015; Pierce, Williams and Martin 2016). Lefebvre and Massey both characterize space as variegated, with multiple processes playing out at once. Yet we think that Lefebvre, with whom we agree in many ways, has sometimes led urban analysts astray. When he writes about the production of space, his work can seem very abstract; when scholars or activists attempt to empirically apply Lefebvre, we see many instances when they try to cleave apart the different "parts" that he insists are a whole, integrated, uncleavable phenomenon (Pierce and Martin 2015).

Where other urban scholars leverage Lefebvre's ideas about how space is produced, we center Doreen Massey's ideas about how place is lived by bundling space. Massey uses place as a vocabulary for describing the "actually existing" lived geographies of people in the world. Both Massey and Lefebvre emphasize the ongoingness of space/place production, and the complex multiplicities of trying to understand these productions. Massey's language of place bundling offers, for us, a more concrete, scalable understanding of place-making; from individuals bundling places cognitively to the always socially mediated time/space trajectories of what we call place framing.

Massey's philosophical work on space and place (1994, 2005) is partially rooted in a feminist critique that highlights how social categories and inequalities are embedded and expressed in place production. The conflation of place with local, home, and homemaking, for example, highlights for Massey the ways that geographical imaginaries draw on and express gendered social norms and structures. This approach and emphasis also opens an opportunity for thinking about urban landscapes from other non-dominant perspectives. The work of sociologists Marcus Anthony Hunter and Zandria Robinson (2018), for example, on *Chocolate Cities*, can be read as highlighting Black American place framing. In one illustration, they argue that the life-trajectory of Aretha Franklin traces the place-making power of gospel and soul music for Black American communities from the Mississippi Delta to industrial cities such as Detroit. A number of geographers working within the subdisciplinary framework of Black geographies also emphasize place-making from the perspective of Black life (Best 2016; Allen, Lawhon, and Pierce 2019; Bledsoe and Wright 2019; Eaves 2017; Hawthorne 2019; Noxolo 2022).

Massey's theoretical work is not inherently focused on cities. Yet, as we noted above, the world is inexorably becoming more and more urbanized with each passing year. It is true that place-making happens everywhere people meet and shape the world, yes; but in cities, where densities are high and competition over different uses of space is constant, place framing is crucial. Place framing is a way of describing contestation over cities that spans everyday disagreements to exceptional ones: from neighborly disagreements over the locations of flower pots to North Korean objections to the everyday character of Seoul.

We believe that place politics is not just the best way to understand place but to understand cities. We've come to feel more strongly that urban production always proceeds through place politics – it is always about place, and it is always incomplete.

In previous writing, we've articulated a kind of practical approach to examining competing place frames that we've modeled here (Pierce, Martin, and Murphy 2011). The first step is to look for place frames; often the most obvious ones reflect dominant discourses about what a city is known for, what people do there, or what makes it special. The second is to probe more deeply into the character and advocacy embedded in a place frame: what perspectives or modes of life are enabled or highlighted by that frame? What landscapes does it celebrate or prioritize (for example, streets for driving or paths for walking)? Recognizing frames and understanding their character involves identifying the actors and institutions who help to articulate the frames, and this third part of place frame analysis also helps to shift focus from a single frame to the ways that actors and institutions often shape and advocate slightly or even radically different place frames. Identifying and interrogating actors enables a shift across frames to see where different perspectives lie and how they conflict. Place frame analysis consists of these four stages: of (1) seeking discourses or characterizations of the city (or some of its parts) or how the city *should be*; (2) examining the dimensions of these frames; and (3) who advocates for them; and (4) how the different frames – and their claimants – intersect and conflict.

The case study chapters of this book each model a place frame analysis. We have identified a place frame about each case and explored its dimensions, then contrasted it with another viewpoint, or frame. The frames we picked each illustrate some of the most dominant narratives about those cities, and then try to shift focus to other modes of life – other ways of thinking about and experiencing places – or to other kinds of actors.

In London, we articulate a distinction between economic and domestic internationalism. These frames are neither fundamentally contradictory nor are they merely unrelated to each other: they express different aspects of relational power in an urban context. Wealth underpins both of these frames, but one is centered on its accumulation while the other is centered on its appropriate use in the home and in the neighborhood. In this way, the two frames reflect different dimensions of urban experience. With regard to the built environment, daily rounds, lifestyle choices, and aesthetics, the two frames differ in what elements they bring into focus and which ones they blur out.

Our discussion of Tehran, written with colleague Azadeh Hadizadeh Esfahani, is similar in some ways to that of London: it identifies frames that articulate both economic priorities and urban life experience, with frames that highlight Islamic nationalism coexisting with a (relatively) liberal cosmopolitanism. Before the revolution, modernism and western assimilation were key elements of nationalist developmentalist discourse. The frame of Islamic nationalism retains this earlier emphasis on large-scale urban infrastructural investment; key modernist tropes have been digested within and recast as part of the Islamist frame. At the same time, its liberal cosmopolitan discourse, which emphasizes social difference and lifestyle choice, persists as a secondary theme, in part enabled by how these more cosmopolitan practices are imbricated with a modernist development agenda. Thus this chapter shows how framing is not only a process that selects physical and social elements as part of a place, but is also a process of iteratively casting meaning upon those elements over time (or "space–time trajectories").

Worcester's frames highlight challenges for smaller cities, in mediating between locally situated decision making and the influences and imperatives of broader scales of action and government. Both its municipal developmentalism, which emphasizes redevelopment of the core downtown infrastructure, and its regional environmentalism, which responded to ecological threats to its treescapes, were shaped and informed by events beyond the city. Specifically, the frames highlight how local economic strategies constantly reference regional dynamics, while regional ecological imperatives faced significant local backlash for being insufficiently attentive to experiences of place at the neighborhood scale. Both of these frames highlight the tensions of local place attachments within a multi-scalar governance context that constantly shifts the scale of action between the municipal level and the regional and federal ones. The narratives in Worcester that emerge point to the capacities of place frames to simultaneously reference multiple geographical scales.

In Portland, we contrast a relatively deracinated "green" urban development framing with a frame that seeks to reinsert that history of racial oppression and, in doing so, to de-emphasize the smug triumphalism of Portland's boosters and admirers. Here, a key part of the framing conflict is who gets admitted to history both ancient and recent. If you see Portland as a development machine for the benefit of its most earnest advocates for an environmentally sensitive urbanism, then the long history of racial exclusion is more easily cast as beyond the mission of the city. Events after the death of George Floyd have made that framing strategy untenable. Because "green" politics in the United States are often situated on the political left, bundling that centers care for the environment and the economically disadvantaged but excludes those who are oppressed through racialization has become increasingly difficult to imagine, and the conflict became explosive.

Our colleague Amy Zhang's discussion of Chongqing points to the ways that place frames also derive significant force from outside of many cities. In one frame she explores, Chongqing's physical landscapes form an iconic imaginary that shapes a tourist place frame about the city, integrating its physical topography with stereotypical understandings of its national context. At the same time, Chongqing's municipal government leverages the distinct geography of the region to innovate urban expansion into rural areas, redefining the city as a site of policy innovation. These two frames together highlight the internal–external tensions of place frames, as people build on the (always imperfect) knowledge of outsiders to promote places, while fighting against the ways that cities are disregarded and reduced to caricatured representations of urbanness.

Our discussion of Jerusalem highlights the impossibilities of fully knowing cities through exploration of the sometimes radically opposed and mutually subjugating dynamics of place frames. Because frames represent different collective agreements on what comprises a city and how it is experienced, some actively subjugate aspects of other frames. These dynamics are not merely representational but express the conflicts and irresolution of some aspects of co-presence in place, always embodying real power differentials between groups. Our case study in this instance (like that of the City of London at the outset) focuses on a geographically small but crucial and contested sphere in Greater Jerusalem: the Old City. We highlight a tourist frame that narrates compatibility of mutual respect among three world religions, yet which in doing so enables a second frame of nationalist Jewish identity. This second frame elides the ongoing domination of a Palestinian people to enable this city's status quo.

Revisiting the three major points of emphasis

In the Introduction we argued that in thinking about cities as places we try to emphasize three things: (1) *the balance between social structures and human agency*; (2) *the incompleteness of urban knowledge*; and (3) *the radical plurality of urban places*. No matter how we look at cities, if we keep these three things in mind as we do so, we can humbly explore the limits of our knowledge and push to try to see cities, and understand them, a little better.

With regard to *human agency*, we emphasize that people are not merely prisoners of the structural processes driving cities. Urban denizens are always acting to shape the urban environment and affect how other people use, plan, and experience it. The idea of human agency here applies equally to someone buying a grand manse in an elite neighborhood, to a renter getting an apartment that is smaller and dingier than they would like because it is all they can afford, or to a person who cannot find or afford a home deciding to sleep under a bridge. In each case, it is clear that economic processes and political decisions far removed from these people's daily lives have provided them with unequal life choices, yet they do make choices within that context.

Geographers and other urban theorists have worked extensively to theorize the degree to which people are governed by structures or free to exercise agency (see box, "Structure and Agency," in chapter 1). An important example of a structure is the relationship between employers and workers: a power-laden relationship that seems natural because it is constantly being reproduced in our late-capitalist economies, yet is ultimately only one possible pattern for relationships around material production among an array of others. Over the past half-century, social scientists have worked to understand the degree to which humans are free to make choices in their lives while enmeshed in structured relationships and processes.

We can recognize that people do not freely and abstractly create their opportunities in life. But in their lives they are making choices, however constrained, every day; and these, too, influence how the city looks and is experienced. Thinking about cities means reflecting on agency, even while understanding it is critically important to focus on systems that structure or create choice because these create the conditions that allow one person to live in a large house while another is trying to figure out the best way to sleep outdoors. But we can sometimes forget that systems and structures intersect with and are shaped by human agents making decisions, including whether and how to question limited choices in the world.

So people, their choices, and activities are important to and for cities. And we want to also examine processes that occur because of non-human things (think, for example, of weather and climate), as well as human systems that operate independently of any one person, such as markets. Indeed, people often refer to "the market" as an independent system with its own logic, despite the fact it is people who set up the rules for how to exchange money and goods through markets – which are often located in urban places. We think the systems or structures that guide and shape the ways that people behave and organize society and space are critically important. But we emphasize agency because a lot of urban theorizing focuses on systems: job and housing markets, transportation arteries, or educational systems, for example. These are all crucial lenses on examining and understanding cities. But they are not complete. Nor are explanations that focus on people's behaviors and choices. They are all informative, and yet also incomplete.

Incompleteness means that there is no settled, agreed-upon city-object: any attempt to examine and understand a city inevitably leaves important elements to the side uninterrogated. Accepting incompleteness as a core urban characteristic may be unsettling because it requires that we acknowledge at the outset that we will fail to fully know or explain cities. But incompleteness can also be, counter-intuitively, an opening for explaining *more*. If we start our thinking about cities by saying we know we haven't yet understood all of the intersecting processes and decisions that shape urban life and structures, then we position ourselves to constantly seek new perspectives and understandings, to try out different ways of thinking.

One of the most powerful ways in which a relational place approach has shaped our own thinking has been in emphasizing that any city is *always* incoherent and unresolved. This is not just to say that people argue over the meaning of a city. It is to say that the constituent parts of "a city" are never fully agreed upon by *anyone*, even those who politically align over partially shared place frames. There is only ever partial agreement. A city is always many cities, many places, all at once.

Portland, OR, for example, has a set of formal municipal boundaries, which may seem to define the city clearly as a place. Actors whose work is defined by those boundaries – such as the mayor or city councilors – recognize the legal entity of incorporated Portland and that their actions as elected officials are bound within that legally incorporated Portland. But these actors also recognize that people drive in and outside of Portland all the time, and the movements in and out of the city – across its legal boundaries – also make Portland, both within and outside of those legal lines, by their social interactions, shopping, paid employment, and experiences of nature.

The legal boundary of Portland doesn't fully define Portland; that formal definition doesn't capture the affection of people for the city's professional basketball team (the Trailblazers) or the fact that major corporate employer Nike is actually based in neighboring Beaverton. The fact that Nike is headquartered in Beaverton also contributes to defining Portland as an urban place, as much as its legal boundaries do. Portland is always many Portlands all at once.

A city never fully coheres. For example, even though some actors believe that defining the formal boundaries of a city categorically defines its spatial limits, legal limits are only one strategy for framing what counts as part of an urban place. All cities are defined multiply, and no one set of actors gets to adjudicate the defining, no matter how well situated institutionally or how much state power they wield. In Portland, many residents see the city's viewshed as "part" of its urban character; others include neighboring municipalities like Beaverton, Tigard, or Milwaukie. Any actor (or group) that frames the city, from an analytical perspective, contributes to the messy overlapping boundary making of the lived city (the place).

On one hand, an analytical focus on the process of place framing rather than on places as produced objects – on the many frames that different place framers employ – foregrounds that there isn't ever a city-as-product that is truly coherent. To study cities using a place-framing approach means abandoning an idea that one can analytically resolve a city. Yet, on the other hand, the idea of a city as a place that never coheres actually emphasizes a key risk in more conventional modes of urban analysis: the dangerous allure of seeing one's analytical lens as more universally applicable than it is.

Consciously thinking about how all city frames are incomplete encourages taking a *radically plural* approach to analysis. Radical pluralism means accepting the incompleteness of our understanding and accounts. It means accepting that more than one explanation might be plausible and even necessary to fully understand something. Indeed, it also means that there is a potentially limitless way of seeing and understanding cities.

Radical plurality suggests two aspects of thinking about cities. The first has to do with what we imagine exists in cities, or the ontology of cities. It means accepting and understanding that there are multiple understandings, perspectives, and dimensions of life and interactions in cities. Cities are, in short, radically plural; they have multiple coexisting and overlapping processes and people. What we think we share with others about our experiences of cities, or even what characterizes a particular city, is actually quite different, even in the same built environment.

Scholars Trevor Barnes and Eric Sheppard have argued for a scholarly stance within their field of economic geography called "engaged pluralism" (Barnes and Sheppard 2010, drawing on Bernstein 1988), which we see as compatible with radical plurality as a way of seeing and understanding cities. Engaged pluralism recognizes multiple ideas and ways of being or understanding something, and it acknowledges that we all hold our own ways of seeing, experiencing, or explaining an issue or a thing. But engaged pluralism requires also being open to someone else's way of seeing, experiencing, and explaining. This approach doesn't mean that both ideas simply coexist; to be engaged, they have to interact. In a city, multiple experiences and ideas and things *do* coexist, and don't always interact. But to understand and explain those things, we need to actively engage urban pluralism in our thinking and theorizing about cities; place framing helps us do that. Barnes and Sheppard (2010: 201) suggest that "pluralism is not useful unless there is engagement among its parties."

Being able to engage across different viewpoints and experiences requires some common concerns, or vocabulary and approach. We cannot simply try to assemble and "hold" in our minds all of the ways to experience and think about cities. In this book, we have used the idea of place framing to offer a vocabulary and approach for thinking about cities. When scholars and people in their everyday lives think about places such as cities (or parts of cities, like neighborhoods), they "frame" them in their minds and in their explanations to emphasize certain things.

Some people refer to economic processes or systems – such as capitalism – to explain the growth and decline of specific areas within cities, the experiences and types of work people find there, and the social and political tensions that play out in cities. For others, cities represent a concatenation of social groups and values, competing for resources or attention. For example, one group of park users might be young people seeking skateboard amenities, who clash with another group of older people who seek a peaceful, quiet refuge in the park. Some people see cities as places where different value systems meet and become weakened by the impersonal marketplace-based logic of interactions, severing the connections people have to familial or religious traditions as they take on behaviors or lifestyles of dominant groups. We want to think about cities as places where all of these things might be happening at once, and we can't a priori decide what is most important to consider in a given place or circumstance.

Why it all matters: place framing makes the effects of everyday urban politics visible and explicit

As we have argued throughout the book, thinking about cities in terms of place framing means rejecting the idea that there is any apolitical or pre-political agreement about what constitutes a given city or a good urban life. Defining "what matters" in a city is always contested. A place-framing approach renders these implicit politics visible. In this way, it displeases all political actors. Some might ask: does this mean that a place-framing approach is simply moral relativism for the city? If everything is political, does that mean there is no "better" or "worse" framing?

We believe that this line of argument misunderstands the goals of place analysis. Admitting that all city making is place framing doesn't mean that one can't have moral, ethical, or practical objections to particular frames or modes of place-making. It also doesn't equate success at framing with any kind of moral authority. Pat Noxolo, considering relational place, asks "whether this additive relationality can really be a sufficiently radical answer for the voraciously exploitative and extractive processes that accompany racial capitalism: poverty, inequality, environmental racism" (2022: 4). Our answer is ultimately: emphatically, yes. Place frames aren't just a set of meanings layered on top of a space that embodies hegemonic power, but they are in fact the way meaning is constituted for all denizens. Place-making is city making: if one seeks to remake cities radically, understanding relational place politics is crucial to making them differently.

Our use of the concepts of space and place builds crucially on work by Doreen Massey, though her writing can be dense and abstract. We find Massey's version of a space/place vocabulary to be non-intuitive but ultimately transformative to our thinking about cities. Massey writes about space and politics: "Conceptualising space as open, multiple and relational, unfinished and always becoming, is a prerequisite for history to be open and thus a prerequisite, too, for the possibility of politics" (Massey 2005: 59).

Politics relies on the notion that change is possible. Usually, we conceptualize time as the operative variable in change; that is, that things will change over the course of time. But Massey is arguing that "things" can't change over time if we aren't also conceptualizing them spatially; life changes when its practice, which is always emplaced, also changes. So being able to imagine and argue for change requires that space and place are as flexible and contingent as time. Space/place and time thus interact and rely upon each other, open to alternative future time/spaces.

Massey uses the language of space here. But recall that, for Massey and for us, place frames are made up of space–time trajectories, not just stories about small areas of a map. When we think about urban politics, we are trying to emphasize that urban places are open, multiple, and relational. Foregrounding place framing as a central activity in urban production helps us see cities in this way.

We as authors believe that cities can be sites of emancipation, both large and small. Political possibility is not, in our view, only evident in formal political arenas, or led by those who have amassed enough power to make large changes to a city or polity. Place framing isn't always small scale, but a key virtue of this approach to thinking about cities is that it is always decentering the most obvious frame in a conflict by bringing it into conversation with other, less empowered efforts to frame a city. It does not *deny* power relations, it *foregrounds* them by demonstrating the many alternative visions of the urban that are rendered invisible by hegemonic urban politics. Thinking of cities as places, and urban production as place-making, is a way of insisting that we bring those stories and efforts into analysis. Whether or not this resonates likely depends in part on the degree to which one sees those other stories as material to the long history of the urban.

Notes

Chapter 2 City of London: A Machine for Living/The Seat of Wealth

1 Though there are several earlier claimants to the throne, the Norman Conquest in 1066 is often used as the historic date of the consolidation of a coherent and consolidated kingdom of England.
2 London was also, of course, the capital of the United Kingdom and thus home to a host of other functional institutions. But these institutions are not centered in the City of London itself, and in fact have never been.
3 Photo by Joseph Pierce.
4 Photo by Joseph Pierce.
5 Charles and Ray Eames were well-known designers for the Herman Miller furniture company in the mid-twentieth century. The Eames chair and the Arco lamp were famous as representative of the mid-century modern aesthetic, and early examples are now coveted for their authenticity (Wilson 2004).

Chapter 3 Tehran: Islamic Developmentalism/Diverse Cosmopolitanism

1 Iran had about 85 million people in total in 2021; Tehran Province is one of 33 provinces in the country.
2 Photo by Azadeh Hadizadeh Esfahani.
3 A well-known slogan of the revolution.

4 The permission to increase the number of floors has been a significant income source for the Tehran municipality for years; in some years it forms 80 percent of the municipality's income. The pressure to create new sources of income started in the 1980s, when the national government decreed that municipalities should be financially self-sufficient, and increased later when some social responsibilities were devolved to Tehran municipality (such as provision of services for homeless people, child laborers, and other needy groups).
5 For example, according to Islamic rules, money kept in bank accounts should not receive any interest, but in practice, and considering the high inflation rate of the country, bank interest does exist and on occasions has reached 22 percent, much higher than bank interest in many non-Islamic societies.
6 A chador is an outer garment that is a full-body-length semicircle of fabric that is open down the front. This cloth is tossed over the woman's head, and she holds it closed in the front. In the formal clothing of the government, women should wear black chadors.
7 A kind of woman's clothing hung from the shoulders to the floor. It should be loose and cover the woman's whole body. The official requirement is to wear long dark mantoos and scarfs. While enforced in governmental and other organizations' settings, this rule is not necessarily enforced on the streets of Tehran.
8 The formal marriage in Iran is legal religious marriage. White marriage is a kind of marriage in which two people live together without any legal religious commitment.
9 In the 1940s, the Soviet Union called for Armenians living in other countries to immigrate back to Armenia, but then reneged on the call. Therefore, many Armenians who were passing through Iran to reach Armenia stayed in Iran and, since then, there are a number of Iranian cities with Armenian populations, including Tehran.
10 Photo by Azadeh Hadizadeh Esfahani.

Chapter 4 Worcester: Local Economic Engine/Regional Forest Under Threat

1 Clark University's Human-Environment Regional Observatory (HERO) program studied the LB infestation and response; presentations about this fieldwork, including details of

interviews, are available at https://www.clarku.edu/departments/hero-program/research/ and in Palmer et al. (2014).
2 Photo by Naomi Shertzer.

Chapter 5 Portland: Paradise of Environmentalism/Legacy of Exclusionary Racism

1 Photo by Ian Sane, Flickr.
2 Photo by joyofresistance, Flickr.

Chapter 6 Chongqing: International Cyberpunk Marvel/National Policy Innovator

1 Photo by Amy Y. Zhang.
2 Photo by David290, Wikimedia Commons.

Chapter 7 Jerusalem: Religious Tourist Destination/Ethno-National Citadel

1 Terms to denote descriptive characteristics of people are often contested and fraught; especially so here (see box, "Toponyms and Demonyms"). Israelis can be any religion. Israel's Declaration of Independence (1948) states clearly that although Israel is a state for Jews, it is also a state for all its inhabitants. Arab citizens of Israel may be Israelis, but they are also simultaneously Palestinians in the sense of an ethnic nationhood made up of Christians, Muslims, and Druze (although many members of the latter group consider themselves Israeli rather than Palestinian). In this chapter, we use "Jewish citizens of Israel" and "Palestinian citizens of Israel" to denote Israeli citizens with a range of religions (recognizing the inherent contradiction in the latter term, and the lack of representation of Druze, who largely live in the northern part of Israel). Some Palestinian citizens of Israel (and most Jewish citizens of Israel) use "Arab Israeli" (Schnell and Haj-Yahya 2014), but the terms we use here seem most parallel to one another. We use "Palestinian non-citizen" for people who do not have Israeli citizenship but live in metropolitan neighborhoods of

Jerusalem, or the West Bank, or Gaza, recognizing that these three sites contain extremely different life experiences and opportunities in relation to Israel and one another. This choice of terms is based on clarifying statuses in relation to this chapter and its focus on Jerusalem, not a statement on broader territorial claims, identities, and relationships overall in the region.
2 Meaning "weeks" and corresponding roughly with the end of the Christian season of Pentecost.
3 A harvest festival, the Festival of Booths, or Tabernacles.
4 Photo by Dan Shertzer.
5 Krauss 2021. See also https://www.btselem.org/jerusalem. Last accessed 4 January 2022. A larger proportion of Palestinian citizens of Israel live in towns and villages, especially in the northern half of Israel in the area of the Sea of Galilee in and around Nazareth, in cities such as Jaffa, Haifa, Akko (Acre), and in the southern Negev region (latter largely Bedouin).
6 One painfully illustrative description of these and other dimensions of Palestinian daily life within Jerusalem and its environs in the West Bank is in Thrall (2021).

Chapter 8 The Impossibilities of Fully Knowing a City

1 https://ourworldindata.org/urbanization (this is based on UN Pop Division data).

References

Adams, P. C., Hoelscher, S., and Till, K. (eds). 2001. *Textures of Place*. Minneapolis, MN: University of Minnesota Press.

Adler, M. 2012. Collective Identity Formation and Collective Action Framing in a Mexican "Movement of Movements." *Interface: A Journal for and About Social Movements* 4(1): 287–315.

Agnew, J. A. 1987. *Place and Politics: The Geographical Mediation of State and Society*. Boston, MA: Allen and Unwin.

Agnew, J. A. 1989. The Devaluation of Place in Social Science, in J. A. Agnew and J. Duncan (eds), *The Power of Place*. Boston, MA: Unwin Hyman, pp. 2–29.

Aiello, G. 2021. Communicating the "World-Class" City: A Visual-Material Approach. *Social Semiotics* 31(1): 136–54.

Al-Haj, M. 1988. The Changing Arab Kinship Structure: The Effect of Modernization in an Urban Community. *Economic Development and Cultural Change* 36(2): 237–58.

Alderman, D. H. 2000. A Street Fit for a King: Naming Places and Commemoration in the American South. *The Professional Geographer* 52(4): 672–84.

Alderman, D. H. 2003. Street Names and the Scaling of Memory: The Politics of Commemorating Martin Luther King, Jr within the African American Community. *Area* 35(2): 163–73.

Alderman, D. H. and Inwood, J. 2013. Street Naming and the Politics of Belonging: Spatial Injustices in the

Toponymic Commemoration of Martin Luther King Jr. *Social & Cultural Geography* 14(2): 211–33.

Allegra, M., Casaglia, A., and Rokem, J. 2012. The Political Geographies of Urban Polarization: A Critical Review of Research on Divided Cities. *Geography Compass* 6(9): 560–74.

Allen, D. L. 2020. Asserting a Black Vision of Race and Place: Florida A&M University's Homecoming as an Affirmative, Transgressive Claim of Place. *Geoforum* 111: 62–72.

Allen, D., Lawhon, M., and Pierce, J. 2019. Placing Race: On the Resonance of Place with Black Geographies. *Progress in Human Geography* 43(6): 1001–19.

Allen, J. and Cochrane, A. 2014. The Urban Unbound: London's Politics and the 2012 Olympic Games. *International Journal of Urban and Regional Research* 38: 1609–24.

Alonso, W. 1960. A Theory of the Urban Land Market. *Papers and Proceedings of the Regional Science Association* 60: 149–57.

Amara, M. and Schnell, I. 2004. Identity Repertoires among Arabs in Israel. *Journal of Ethnic and Migration Studies* 30(1): 175–93.

Amin, A. 2002. Ethnicity and the Multicultural City: Living with Diversity. *Environment and Planning A* 34(6): 959–80.

Anand, N. 2017. *Hydraulic City: Water and the Infrastructures of Citizenship in Mumbai.* Durham, NC: Duke University Press.

Anguelovski, I., Connolly, J. J., Garcia-Lamarca, M., Cole, H., and Pearsall, H. 2019. New Scholarly Pathways on Green Gentrification: What Does the Urban "Green Turn" Mean and Where Is It Going? *Progress in Human Geography* 43(6): 1064–86.

Aoki, K. 1993. Race, Space, and Place: The Relation between Architectural Modernism, Post-Modernism, Urban Planning, and Gentrification. *Fordham Urban Law Journal* 20(40): 699–829.

Arreola, D. D. 1988. Mexican American Housescapes. *Geographical Review* 78(3): 299–315.

Asafu-Adjaye, J., Blomquist, L., Brand, S., et al. 2015. *An*

Ecomodernist Manifesto. Accessed online July 20, 2022 at https://www.ecomodernism.org/.

Avila, E. 2014. *The Folklore of the Freeway*. Minneapolis, MN: University of Minnesota Press.

Azaryahu, M. and Golan, A. 2001. (Re)naming the Landscape: The Formation of the Hebrew Map of Israel 1949–1960. *Journal of Historical Geography* 27(2): 178–95.

Bakker, K. 2010. *Privatizing Water: Governance Failure and the World's Urban Water Crisis*. Ithaca, NY: Cornell University Press.

Bakker, K. 2012. Water: Political, Biopolitical, Material. *Social Studies of Science* 42(4): 616–23.

Ballou, B. 2018. PawSox Move to Worcester Started with Gene Zabinski's Postcard Campaign. *Telegram and Gazette* (Worcester, MA), 18 August. Accessed online June 22, 2022 at https://eu.telegram.com/story/sports/mlb/woosox/2018/08/18/pawsox-move-to-worcester-started-with-gene-zabinskis-postcard-campaign/11000182007/.

Barnes, T. J. and Sheppard, E. 2010. "Nothing Includes Everything": Towards Engaged Pluralism in Anglophone Economic Geography. *Progress in Human Geography* 34(2): 193–214.

Baudrillard, J. 1988. *America*. London: Verso.

Beal, V. and Pinson, G. 2014. When Mayors Go Global: International Strategies, Urban Governance and Leadership. *International Journal of Urban and Regional Research* 38: 302–17.

Benford, R. D. 1993. Frame Disputes within the Nuclear Disarmament Movement. *Social Forces* 71(3): 677–701.

Berg, L. and Kearns, R. 1996. Naming as Norming: "Race", Gender, and the Identity Politics of Naming Places in Aotearoa/New Zealand. *Environment and Planning D: Society and Space* 14: 99–122.

Bernstein, R. J. 1988. Pragmatism, Pluralism and the Healing of Wounds. *Proceedings and Addresses of the American Philosophical Association* 63: 5–18.

Best, A. 2016. The Way They Blow the Horn: Caribbean Dollar Cabs and Subaltern Mobilities. *Annals of the American Association of Geographers* 106(2): 442–9.

Bledsoe, A. and Wright, W. J. 2019. The Pluralities of Black Geographies. *Antipode* 51(2): 419–37.

Blomley, N. 2003. *Unsettling the City: Urban Land and the Politics of Property*. New York: Routledge.

Borgonjon, D. X. 2018. Continental Drift: Notes on "Asian" Art. *Rhizome*. Accessed online June 10, 2022 at https://rhizome.org/editorial/2018/sep/05/continental-drift-notes-on-asian-art/.

Bourdieu, P. 1977. *Outline of a Theory of Practice*. Oxford: Oxford University Press.

Brand, S. 2009. *Whole Earth Discipline: An Ecopragmatist Manifesto*. New York: Viking Penguin.

Brenner, N. 2000. The Urban Question as a Scale Question: Reflections on Henri Lefebvre, Urban Theory and the Politics of Scale. *International Journal of Urban and Regional Research* 24(2): 363–78.

Brenner, N. and Schmid, C. 2015. Towards a New Epistemology of the Urban? *City* 19(2–3): 151–82.

Broadbent, J. 1989. Strategies and Structural Contradictions: Growth Coalition Politics in Japan. *American Sociological Review* 54(5): 707–21.

Bruce, B. 2019. The Rise and Fall of the Ku Klux Klan in Oregon during the 1920s. *Voces Novae* (Chapman University): 11(2). Accessed online May 5, 2022 at https://digitalcommons.chapman.edu/vocesnovae/vol11/iss1/2/.

Bullard, R. D. (ed.). 1993. *Confronting Environmental Racism: Voices from the Grassroots*. Boston, MA: South End Press.

Carson, R., with Wilson, E. O. 2002 [1962]. *Silent Spring*. United Kingdom: Houghton Mifflin.

Castells, M. 1977 [1972]. *The Urban Question*. Cambridge, MA: MIT Press.

Castells, M. 1983. *The City and the Grassroots*. London: Edward Arnold.

Chan, D. 2016. Asia-futurism. *Artforum* 54(10). Accessed online June 10, 2022 at https://www.artforum.com/print/201606/asia-futurism-60088/.

Chan, K. W. 2010. The Global Financial Crisis and Migrant Workers in China: "There is No Future as a Labourer;

Returning to the Village has No Meaning". *International Journal of Urban and Regional Research* 34(3): 659–77.

Chan, K. W. and Buckingham, W. 2008. Is China Abolishing the Hukou System? *China Quarterly* 195: 582–606.

Chan, K. W. and Wei, Y. 2019. Two Systems in One Country: The Origin, Functions, and Mechanisms of the Rural–Urban Dual System in China. *Eurasian Geography and Economics* 60(4): 422–54.

Chan, K. W. and Zhang, L. 1999. The Hukou System and Rural–Urban Migration in China: Processes and Changes. *China Quarterly* 160: 818–55.

Chauncey, G. 2008 [1994]. *Gay New York: Gender, Urban Culture, and the Making of the Gay Male World, 1890–1940*. New York: Basic Books.

Christophers, B. 2016. For Real: Land as Capital and Commodity. *Transactions of the Institute of British Geographers* 41(2): 134–48.

Christophers, B. 2019. Putting Financialisation in its Financial Context: Transformations in Local Government-led Urban Development in Post-financial Crisis England. *Transactions of the Institute of British Geographers* 44(3): 571–86.

Chun, W. H. K. 2006. *Control and Freedom: Power and Paranoia in the Age of Fiber Optics*. Cambridge, MA: MIT Press.

Clark, W. A. V. 1991. Residential Preferences and Neighborhood Racial Segregation: A Test of the Schelling Segregation Model. *Demography* 28(1): 1–19.

Clarke, N. 2012. Urban Policy Mobility, Anti-politics, and Histories of the Transnational Municipal Movement. *Progress in Human Geography* 35(4): 1–19.

Cohen, M. 2017. A Systematic Review of Urban Sustainability Assessment Literature. *Sustainability* 9(11): 2048.

Collins, F. L. 2012. Transnational Mobilities and Urban Spatialities: Notes from the Asia-Pacific. *Progress in Human Geography* 36(3): 316–35.

Cresswell, T. 2014. *Place: A Short Introduction*, 2nd edn. Hoboken, NJ: Wiley Publishing.

Cresswell, T. 2019. *Maxwell Street: Writing and Thinking Place*. Chicago, IL: University of Chicago Press.

Cronon, W. 1992. *Nature's Metropolis: Chicago and the Great West*. London: W. W. Norton.

Curran, W. and Hamilton, T. (eds). 2017. *Just Green Enough: Urban Development and Environmental Gentrification*. Abingdon: Routledge.

Da Cruz, N. F. and Marques, R. C. 2011. Viability of Municipal Companies in the Provision of Urban Infrastructure Services. *Local Government Studies* 37(1): 93–110.

Davidson, A. 2016. What Happened to Worcester? *New York Times Magazine*, 27 April. Accessed online December 3, 2020 at https://www.nytimes.com/2016/05/01/magazine/what-happened-to-worcester.html/.

Davidson, M. 2009. Displacement, Space and Dwelling: Placing Gentrification Debate. *Ethics Place and Environment* 12(2): 219–34.

Davidson, M., Lukens, D., and Ward, K. 2021. The Post-Great Recession Geographies of US Municipal Borrowing and Indebtedness. *Professional Geographer* 73(2): 240–53.

de Seta, G. 2020. Sinofuturism as Inverse Orientalism: China's Future and the Denial of Coevalness. *SFRA Review* 50(2–3): 86–94.

de Vries, S., Verheij, R. A., Groenewegen, P. P., and Spreeuwenberg, P. 2003. Natural Environments – Healthy Environments? An Exploratory Analysis of the Relationship between Greenspace and Health. *Environment and Planning A* 35: 1717–31.

DeFilippis, J. and Wyly, E. 2008. Running to Stand Still: Through the Looking Glass with Federally Subsidized Housing in New York City. *Urban Affairs Review* 43(6): 777–816.

Denis, J. and Pontille, D. 2021. Maintenance Epistemology and Public Order: Removing Graffiti in Paris. *Social Studies of Science* 51(2): 233–58.

Devine-Wright, P. 2015. Local Attachments and Identities: A Theoretical and Empirical Project across Disciplinary Boundaries. *Progress in Human Geography* 39 (4): 527–30.

Ding, C. R. 2004. Urban Spatial Development in the Land

Policy Reform Era: Evidence from Beijing. *Urban Studies* 41(10): 1889–1907.
Donovan, G. H. and Butry, D. T. 2009. The Value of Shade: Estimating the Effect of Urban Trees on Summertime Electricity Use. *Energy and Buildings* 41(6): 662–68.
Du Bois, W. E. B. 2014 [1899]. *The Philadelphia Negro: A Social Study*. Oxford: Oxford University Press.
Duneier, M. 2016. *Ghetto: The Invention of a Place, the History of an Idea*. New York: Farrar, Straus, and Giroux.
Durkheim, E. 1997 [1893]. *The Division of Labor in Society*, trans. W. D. Halls. New York: Simon and Schuster.
Dwyer, O. J. and Alderman, D. H. 2008. Memorial Landscapes: Analytic Questions and Metaphors. *GeoJournal* 73(3): 165–78.
Easley, G. 1992. *Staying Inside the Lines: Urban Growth Boundaries*. Chicago, IL: American Planning Association.
Eaton, W. M., Gasteyer, S. P., and Busch, L. 2014. Bioenergy Futures: Framing Sociotechnical Imaginaries in Local Places. *Rural Sociology* 79(2): 227–56.
Eaves, L. E. 2017. Black Geographic Possibilities: On a Queer Black South. *Southeastern Geographer* 57(1): 80–95.
Ehrkamp, P. 2005. Placing Identities: Transnational Practices and Local Attachments of Turkish Immigrants in Germany. *Journal of Ethnic and Migration Studies* 31(2): 345–64.
Ehrlich, P. R. 1968. *The Population Bomb*. New York: Ballantine.
Elmes, A, Healy, M., Geron, N., et al. 2020. Mapping Spatiotemporal Variability of the Urban Heat Island across an Urban Gradient in Worcester, Massachusetts Using in Situ Thermochrons and Landsat-8 Thermal Infrared Sensor (TIRS) Data. *GIScience & Remote Sensing* 57(7): 845–64.
Elwood, S. 2006. Beyond Cooptation or Resistance: Urban Spatial Politics, Community Organizations, and GIS-based Spatial Narratives. *Annals of the Association of American Geographers* 96(2): 323–41.
Engels, F. 2010 [1845]. *The Condition of the Working-Class in England in 1844*. Cambridge, UK: Cambridge University Press.

Evans, G. 2003. Hard-branding the Cultural City – from Prado to Prada. *Urban Studies* 27(2): 417–40.

Fan, C. C. 2008. *China on the Move: Migration, the State, and the Household*. London: Routledge.

Fang, C. 2015. Scientifically Selecting and Hierarchically Nurturing China's Urban Agglomerations for the New Normal. *Bulletin of the Chinese Academy of Sciences* 30(2): 127–36.

Fang, C. and Yu, D. 2017. Urban Agglomeration: An Evolving Concept of an Emerging Phenomenon. *Landscape and Urban Planning* 162: 126–36.

Florida, R. 2002. *The Rise of the Creative Class*. New York: Basic Books.

Florida, R. 2014. *The Rise of the Creative Class – Revisited: Revised and Expanded*. New York: Basic Books.

Fominaya, C. F. 2010. Creating Cohesion from Diversity: The Challenge of Collective Identity Formation in the Global Justice Movement. *Sociological Inquiry* 80(3): 377–404.

Gamson W. A. 1992. *Talking Politics*. Cambridge, UK: Cambridge University Press.

Gandy, M. 2002. *Concrete and Clay: Reworking Nature in New York City*. Cambridge, MA : MIT Press.

Gibson-Graham, J. 1996. *End of Capitalism (as We Knew It)?* Oxford: Blackwell Publishers.

Giddens, A. 1976. *New Rules of Sociological Method*. London, UK: Hutchinson.

Giddens, A. 1984. *The Constitution of Society: Outline of the Theory of Structuration*. United Kingdom: University of California Press.

Gillham, O. 2002. *The Limitless City: A Primer on the Urban Sprawl Debate*. Washington, DC: Island Press.

Gleeson, B. and Spiller, M. 2012. Metropolitan Governance in the Urban Age: Trends and Questions. *Current Opinion in Environmental Sustainability* 4(4): 393–7.

Goffman, E. 1974. *Frame Analysis: An Essay on the Organization of Experience*. Cambridge, MA: Harvard University Press.

Gold, J. 2007. *The Practice of Modernism: Modern Architects* and Urban Transformation, 1954–1972. New York: Taylor & Francis.

Gomez, M. V. 1998. Reflective Images: The Case of Urban Regeneration in Glasgow and Bilbao. *International Journal of Urban and Regional Research* 22(1): 106–21.

Gottmann, J. 1957. Megalopolis or the Urbanization of the Northeastern Seaboard. *Economic Geography* 33(3): 189–200.

Griffiths, R. 1998. Making Sameness: Place Marketing and the New Urban Entrepreneurialism, in N. Oatley (ed.), *Cities, Economic Competition and Urban Policy*. London: Paul Chapman Publishing, pp. 41–57.

Gu, C., Chai, Y., and Cai, J. 1999. *Urban Geography in China*. Beijing: Commercial Press.

Hackworth, J. and Smith, N. 2001. The Changing State of Gentrification. *Tijdschrift voor economische en sociale geografie* 92(4): 464–77.

Hall, P. 2014. *Cities of Tomorrow: An Intellectual History of Urban Planning and Design since 1880*. Chichester: Wiley Blackwell.

Harris, A. 2011. Branding Hoxton: Cultural Landscapes of Postindustrial London, in A. Pike (ed.), *Brands and Branding Geographies*. Cheltenham: Edward Elgar, pp. 187–99.

Harvey, D. 1978. The Urban Process under Capitalism: A Framework for Analysis. *International Journal of Urban and Regional Research* 2(1–3): 101–31.

Harvey, D. 1989a. From Managerialism to Entrepreneurialism: The Transformation in Urban Governance in Late Capitalism. *Geografiska Annaler Series B Human Geography* 71: 3–17.

Harvey, D. 1989b. *The Urban Experience*. Oxford: Blackwell.

Harvey, D. 1990. *The Condition of Postmodernity*. Oxford: Blackwell.

Harvey, D. 1996. *Justice, Nature and the Geography of Difference*. Hoboken, NJ: Wiley-Blackwell.

Harvey, D. 2012. *Rebel Cities: From the Right to the City to the Urban Revolution*. London: Verso.

Hawthorne, C. 2019. Black Matters are Spatial Matters: Black

Geographies for the Twenty-first Century. *Geography Compass* 13(11). DOI: 10.1111/gec3.12468.

Hazboun, S. O., Briscoe, M., Givens, J., and Krannich, R. 2019. Keep Quiet on Climate: Assessing Public Response to Seven Renewable Energy Frames in the Western United States. *Energy Research & Social Science* 57: 101–43.

He, C., Pan, F., and Yan, Y. 2012. Is Economic Transition Harmful to China's Urban Environment? Evidence from Industrial Air Pollution in Chinese Cities. *Urban Studies* 49(8): 1767–90.

Heynen, N., Perkins, H. A., and Roy, P., 2006. The Political Ecology of Uneven Urban Green Space: The Impact of Political Economy on Race and Ethnicity in Producing Environmental Inequality in Milwaukee. *Urban Affairs Review* 42(1): 3–25.

Hoekstra, M. S. 2018. Governing Difference in the City: Urban Imaginaries and the Policy Practice of Migrant Incorporation. *Territory, Politics, Governance* 6(3): 362–80.

Holloway, L. and Hubbard, P. 2001. *People and Place: The Extraordinary Geographies of Everyday Life*. London: Routledge.

Hsing, Y.-T. 2010. *The Great Urban Transformation: Politics of Land and Property in China*. Oxford: Oxford University Press.

Huang, P. C. C. 2011. Chongqing: Equitable Development Driven by a "Third Hand"? *Modern China* 37(6): 569–622.

Huber, M. T. and Currie, T. M., 2007. The Urbanization of an Idea: Imagining Nature through Urban Growth Boundary Policy in Portland, Oregon. *Urban Geography* 28(8): 705–31.

Hume, S. E. 2015. Two Decades of Bosnian Place-making in St Louis, Missouri. *Journal of Cultural Geography* 32(1): 1–22.

Hunt, S. A., Benford, R. D., and Snow, D. A. 1994. Identity Fields: Framing Processes and the Social Construction of Movement Identities, in E., Laraña, H. Johnston, and J. R. Gusfield (eds), *New Social Movements: From Ideology to Identity*. Philadelphia, PA: Temple University Press, pp. 185–208.

Hunter, M. A. 2013. *Black Citymakers*. Oxford: Oxford University Press.

Hunter, M. A. and Robinson, Z. F. 2018. *Chocolate Cities: The Black Map of American Life*. Berkeley, CA: University of California Press.

Ibes, D. C. 2016. Integrating Ecosystem Services into Urban Park Planning & Design. *Cities and the Environment (CATE)* 9(1): Article 1.

Inwood J. F. J. and Martin, D. G. 2008. Whitewash: White Privilege and Racialized Landscapes at the University of Georgia. *Social & Cultural Geography* 9(4): 373–95.

Irwin, B. 2019. Abstract City: The Phenomenological Basis for the Failures of Modernist Urban Design. *Journal of Aesthetics and Phenomenology* 6(1): 41–58.

Iveson, K. 2013. Cities within the City: Do-it-yourself Urbanism and the Right to the City. *International Journal of Urban and Regional Research* 37(3): 941–56.

Jackson, K. T. 1985. *Crabgrass Frontier: The Suburbanization of the United States*. New York: Oxford University Press.

Jiang, Y. and Waley, P. 2020. Who Builds Cities in China? How Urban Investment and Development Companies Have Transformed Shanghai. *International Journal of Urban and Regional Research* 44(4): 636–51.

Johnson, L. 2017. Bordering Shanghai: China's Hukou System and Processes of Urban Bordering. *Geoforum* 80: 93–102.

Joo, Y. M. and Hoon Park, S. 2017. Overcoming Urban Growth Coalition: The Case of Culture-led Urban Revitalization in Busan, South Korea. *Urban Affairs Review*. 53(5): 843–67.

Kaika, M. 2010. Architecture and Crisis: Re-inventing the Icon, Re-imag(in)ing London and Re-branding the City. *Transactions of the Institute of British Geographers* NS 35: 453–74.

Kaplan, D. H. 2015. Immigration and the Making of Place in Paris. *Journal of Cultural Geography* 32(1): 23–39.

Kaplan, D. H. and Recoquillon, C. 2014. Ethnic Place Identity within a Parisian Neighborhood. *Geographical Review* 104(1): 33–51.

Kavaratzis, M. and Ashworth, G. J. 2005. City Branding: An

Effective Assertion of Identity or a Transitory Marketing Trick? *Tijdschrift voor Economische en Sociale Geografie* 96(5): 506–14.

Kean, F. L. 2019. *On Shifting Foundations: State Rescaling, Policy Experimentation and Economic Restructuring in Post-1949 China.* Chichester, UK: Wiley.

Kirby, A., Knox, P., and Pinch, S. (eds). 2017. *Public Service Provision and Urban Development.* Abingdon, UK: Taylor & Francis.

Klasander, A. 2005. Challenges of the Modernist Urban Landscape: On Urban Design and (Sub)Urban Space. *Nordisk Arkitekturforskning* 2005(1): 37–46.

Kotsopoulos, N. 2020. Tax Deal Revisited as Unum Misses Jobs Benchmark. *Telegram and Gazette* (Worcester, MA), 8 November. Accessed online November 17, 2020 at https://www.telegram.com/story/news/2020/11/08/unums-tax-deal-worcester-revised-insurer-misses-jobs-benchmark/6213246002/.

Krauss, J. 2021. Palestinians Feel Loss of Family Homes as Evictions Loom. AP News, 10 May. Accessed online January 4, 2022 at https://apnews.com/article/middle-east-religion-2ba6f064df3964ceafb6e2ff02303d41/.

Kurtz, H. E. 2003. Scale Frames and Counter-scale Frames: Constructing the Problem of Environmental Injustice. *Political Geography* 22(8): 887–916.

Lang, J. 2018. Chongqing's 24-Storey "Wanghong Building" Has No Elevator; Three Exits Lead to Three Different Streets. *Xinhua Net*, 19 January. Accessed (in Chinese) online August 12, 2022 at http://www.xinhuanet.com/politics/2018-01/19/c_1122281315.htm/.

Larsen, S. C. 2004. Place, Activism, and Development Politics in the Southwest Georgia United Empowerment Zone. *Journal of Cultural Geography* 22(1): 27–49.

Lauermann, J. 2016. Boston's Olympic Bid and the Evolving Urban Politics of Event-led Development. *Urban Geography* 37(2): 313–21.

Lauermann, J. 2018. Municipal Statecraft: Revisiting the Geographies of the Entrepreneurial City. *Progress in Human Geography* 42(2): 205–24.

Le Corbusier. 1927. *Towards a New Architecture*, trans. Frederick Etchells. London: John Rodker.
Lee, S. Y. and Han, Y. 2020. When Art Meets Monsters: Mapping Art Activism and Anti-gentrification Movements in Seoul. *City, Culture and Society* 21: 100292.
Lees, L., Slater, T., and Wyly, E. 2013. *Gentrification*. Abingdon, UK: Routledge.
Lefebvre, H. 1991 (1974). *The Production of Space*, trans. David Nicholson Smith. Malden, MA: Blackwell.
Lefebvre, H. 1996. *Writings on Cities*, trans. and ed. Eleonore Kofman and Elizabeth Lebas. Oxford: Basil Blackwell.
Leib, J. I. 2002. Separate Times, Shared Spaces: Arthur Ashe, Monument Avenue and the Politics of Richmond, Virginia's Symbolic Landscape. *Cultural Geographies* 9(3): 286–312.
Leitner, H. 1990. Cities in Pursuit of Economic Growth: The Local State as Entrepreneur. *Political Geography Quarterly* 9(2): 146–70.
Lewicka, M. 2008. Place Attachment, Place Identity, and Place Memory: Restoring the Forgotten City Past. *Journal of Environmental Psychology* 28(3): 209–31.
Lewis, P. 1979. Axioms for Reading the Landscape, in D. W. Meinig (ed.), *The Interpretation of Ordinary Landscapes*. New York: Oxford University Press, pp. 11–32.
Lin, G. C. S. and Ho, S. P. S. 2005. The State, Land System, and Land Development Processes in Contemporary China. *Annals of the Association of American Geographers* 95(2): 411–36.
Lin, G. C. S. and Zhang, A. Y. 2017. China's Metropolises in Transformation: Neoliberalizing Politics, Land Commodification, and Uneven Development in Beijing. *Urban Geography* 38(5): 643–65.
Liu, Y. and Lin, G. C. S. 1998. Changing Central-local Relation in Post-reform China: A Geographical Perspective. *Journal of Chinese Geography* 8(3): 203–20.
Locke, D. H., Hall, B., Grove, J. M., et al. 2021. Residential Housing Segregation and Urban Tree Canopy in 37 US Cities. *npj Urban Sustainability* 1(1): 1–9.
Logan, J. and Molotch, H. 1987. *Urban Fortunes: The Political Economy of Place*.

MacLeod, G. 2011. Urban Politics Reconsidered: Growth Machine to Post-democratic City? *Urban Studies* 48: 2629–60.

Main, K. and Sandoval, G. F. 2015. Placemaking in a Translocal Receiving Community: The Relevance of Place to Identity and Agency. *Urban Studies* 52(1): 71–86.

Marston, S. A. 2002. Making Difference: Conflict over Irish Identity in the New York City St Patrick's Day Parade. *Political Geography* 21(3): 373–92.

Martin, D. G. 2003a. Enacting Neighborhood. *Urban Geography* 24(5): 361–85.

Martin, D. 2003b. "Place-framing" as Place-making: Constituting a Neighborhood for Organizing and Activism. *Annals of the Association of American Geographers* 93(3): 730–50.

Martin, D. G. 2004. Nonprofit Foundations and Grassroots Organizing: Reshaping Urban Governance. *Professional Geographer* 56(3): 394–405.

Martin, D. G. and Miller, B. 2003. Space and Contentious Politics. *Mobilization: An International Journal* 8(2): 143–56.

Martin, D. G. and Scherr, A. 2005. Lawyering Landscapes: Lawyers as Constituents of Landscape. *Landscape Research* 30(3): 379–93.

Martinez, M. H. and Cartier, C. 2017. City as Province in China: The Territorial Urbanization of Chongqing. *Eurasian Geography and Economics* 58(2): 201–30.

Massey, D. 1984. *Spatial Divisions of Labour: Social Structures and the Geography of Production*. Basingstoke, UK: Macmillan.

Massey, D. 1991. A Global Sense of Place. *Marxism Today*. London: Arnold.

Massey, D. 1994. *Space, Place, and Gender*. Cambridge, UK: Polity Press.

Massey, D. 2005. *For Space*. London: Sage Publications.

Massey, D. and Catalano, A. 1978. *Capital and Land: Landownership by Capital in Great Britain*. London: Edward Arnold.

Massey, D. S. and Denton, N. 1993. *American Apartheid:*

Segregation and the Making of the Underclass. Cambridge, MA: Harvard University Press.

McCann E. J. 2008. Expertise, Truth, and Urban Policy Mobilities: Global Circuits of Knowledge in the Development of Vancouver, Canada's "Four Pillars" Drug Strategy. *Environment and Planning A* 40: 885–904.

McCann, E. J. 2011. Points of Reference: Knowledge of Elsewhere in the Politics of Urban Drug Policy, in E. J. McCann and K. Ward (eds), *Mobile Urbanism: Cities and Policymaking in the Global Age*. Minneapolis, MN: University of Minnesota Press, pp. 97–121.

McCann, E. J. and Ward, K. 2011. *Mobile Urbanism: Cities and Policymaking in the Global Age*. Minneapolis, MN: University of Minnesota Press.

McFarlane, C. 2009. Translocal Assemblages: Space, Power and Social Movements. *Geoforum* 40(4): 561–7.

McFarlane, C. 2012. The Entrepreneurial Slum: Civil Society, Mobility and the Co-production of Urban Development. *Urban Studies* 49: 2795–816.

McKittrick, K. 2006. *Black Women and the Cartographies of Struggle*. Minneapolis, MN: University of Minnesota Press.

McKittrick, K. and Woods, C. (eds). 2007. *Black Geographies and the Politics of Place*. Boston: South End Press.

McMorrow, P. 2013. The Comeback of Worcester's Downtown. *Boston Globe*, 10 December. Accessed online November 24, 2020 at https://www.bostonglobe.com/opinion/2013/12/10/worcester-turn-around/l09XCoLmvidqHSvHwEfhgO/story.html/.

McNeill, D. 2011. Airports, Territoriality, and Urban Governance, in E. J. McCann and K. Ward (eds), *Mobile Urbanism: Cities and Policymaking in the Global Age*. Minneapolis, MN: University of Minnesota Press.

Mitchell, D., 1997. The Annihilation of Space by Law: The Roots and Implications of Anti-Homeless Laws in the United States. *Antipode* 29(3): 303–35.

Mitchell, D. 2003. Cultural Landscapes: Just Landscapes or Landscapes of Justice? *Progress in Human Geography* 27(6): 787–96.

Mitnick, J. 2015. For Some Palestinians in East Jerusalem, a Pragmatic Israelification, 6 July. Accessed online January 4, 2022 at https://www.csmonitor.com/World/Middle-East/2015/0706/For-some-Palestinians-in-East-Jerusalem-a-pragmatic-Israelification/.

Modai-Snir, T. and van Ham, M. 2018. Neighbourhood Change and Spatial Polarization: The Roles of Increasing Inequality and Divergent Urban Development. *Cities* 82: 108–18.

Molotch, H. 1976. The City as a Growth Machine: Toward a Political Economy of Place. *American Journal of Sociology* 82: 309–32.

Morley, D. and Robins, K. 1995. *Spaces of Identity: Global Media, Electronic Landscapes and Cultural Boundaries.* London: Routledge.

Morris, A. 2015. *The Scholar Denied: W. E. B. Du Bois and the Birth of Modern Sociology.* Berkeley and Los Angeles, CA: University of California Press.

Muir, J. 1901. *Our National Parks.* Boston, MA: Houghton, Mifflin.

Müller, M. 2017. How Mega-events Capture their Hosts: Event Seizure and the World Cup 2018 in Russia. *Urban Geography* 38(8): 1113–32.

Müller, N., Murray, I. and Blázquez-Salom, M. 2021. Short-term Rentals and the Rentier Growth Coalition in Pollença (Majorca). *Environment and Planning A: Economy and Space* 53(7): 1609–29.

Murphy, J. T. 2015. Human Geography and Socio-technical Transition Studies: Promising Intersections. *Environmental Innovation and Societal Transitions* 17: 73–91.

Nath, T. K., Han, S. S. Z., and Lechner, A. M. 2018. Urban Green Space and Well-being in Kuala Lumpur, Malaysia. *Urban Forestry & Urban Greening* 36: 34–41.

Nowak, D. J., Crane, D. E., and Stevens, J. C. 2006. Air Pollution Removal by Urban Trees and Shrubs in the United States. *Urban Forestry & Urban Greening* 4(3–4): 115–23.

Noxolo, P. 2022. Geographies of Race and Ethnicity 1: Black Geographies. *Progress in Human Geography.*

Accessed online July 22, 2022 at https://doi.org/10.1177/03091325221085291.
Olmsted, F. L. 1870. Public Parks and the Enlargement of Towns, in C. E. Beveridge, C. F. Hoffman, and K. Hawkins (eds), *The Papers of Frederick Law Olmsted. Supplementary Series 1: Writing on Public Parks, Parkways and Park Systems*. Baltimore, MD: Johns Hopkins University Press, pp. 171–205.
Omi, M. and Winant, H. 2014. *Racial Formation in the United States*. New York: Taylor & Francis.
ONS (Office of National Statistics). 2021. Gross Disposable Housing Incomes. Accessed online January 20, 2022 at https://www.ons.gov.uk/economy/regionalaccounts/grossdisposablehouseholdincome/datasets/regionalgrossdisposablehouseholdincomebycombinedauthorityandcityregionsoftheuk/.
Palmer, T. C. 2004. Worcester, Mass., Officials, Investors Hope $300 Million Plan Sparks Revival. *Boston Globe*. Accessed online June 22, 2020 at http://archive.boston.com/business/articles/2004/06/30/waking_up_worcester/.
Palmer, S., Martin, D., DeLauer, V., and Rogan, J. 2014. Vulnerability and Adaptive Capacity in the Asian Longhorned Beetle Infestation in Worcester, Massachusetts. *Human Ecology* 42(6): 965–77.
Palta, M. M., Grimm, N. B., and Groffman, P. M. 2017. "Accidental" Urban Wetlands: Ecosystem Functions in Unexpected Places. *Frontiers in Ecology and the Environment* 15(5): 248–56.
Pan, F. H., Zhang, F. M., Zhu, S. J., and Wójcik, D. 2017. Developing by Borrowing? Inter-jurisdictional Competition, Land Finance and Local Debt Accumulation in China. *Urban Studies* 54(4): 897–916.
Park, R. E., Burgess, E. W., and McKenzie, R. D. 1925. *The City*. Chicago, IL: University of Chicago Press.
Peck, J. 2011. Geographies of Policy: From Transfer-Diffusion to Mobility-Mutation. *Progress in Human Geography* 35(6): 773–97.
Pierce, J. 2011. Urban Land Tenure and Sustainable Practices

in Portland, Oregon. Worcester, MA: Clark University, PhD diss.
Pierce, J. 2022. How Can We Share Space? Ontologies of Spatial Pluralism in Lefebvre, Butler, and Massey. *Space and Culture* 25(1): 20–32.
Pierce, J. and Martin, D. G. 2015. Placing Lefebvre. *Antipode* 47(5): 1279–99.
Pierce, J., Martin, D. G., and Murphy, J. T. 2011. Relational Place-making: The Networked Politics of Place. *Transactions of the Institute of British Geographers* 36(1): 54–70.
Pierce, J., Williams, O., and Martin, D. G. 2016. Rights in Places: An Analytical Extension of the Right to the City. *Geoforum* 70: 79–88
Pinder, D. 2008. Urban Interventions: Art, Politics and Pedagogy. *International Journal of Urban and Regional Research* 32(3): 730–6.
Polletta, F. and Jasper, J. M. 2001. Collective Identity and Social Movements. *Annual Review of Sociology* 27(1): 283–305.
Ponder, C. S. and Omstedt, M. 2019. The Violence of Municipal Debt: From Interest Rate Swaps to Racialized Harm in the Detroit Water Crisis. *Geoforum* 132: 271–80.
Pouyat, R. V., Yesilonis, I. D., Dombos, M., et al. 2015. Global Comparison of Surface Soil Characteristics across Five Cities. *Soil Science* 180(4–5): 136–45.
Pulido, L. 2000. Rethinking Environmental Racism: White Privilege and Urban Development in Southern California. *Annals of the Association of American Geographers* 90(1): 12–40.
Pulido, L. 2016. Flint, Environmental Racism, and Racial Capitalism. *Capitalism, Nature, Socialism* 27(3): 1–16.
Relph, E. 1976. *Place and Placelessness*. London: Pion.
Roast, A. 2019. A Letter from Chongqing. *Tribune Magazine*. Accessed online July 22, 2022 at https://tribunemag.co.uk/2019/05/a-letter-from-chongqing/.
Roast, A. 2021. Three Theses on the Sinofuturist City. *Verge: Studies in Global Asias* 7(2): 80–6.
Sinofuturist City. *Verge: Studies in Global Asias* 7(2): 80–6.

Roberts, K. D. 1997. China's "Tidal Wave" of Migrant Labor: What Can We Learn from Mexican Undocumented Migration to the United States? *International Migration Review* 31(2): 249–93.

Robinson, J. 2002. Global and World Cities: A View from Off the Map. *International Journal of Urban and Regional Research* 26(3): 531–54.

Robinson, Z. F. 2014. *This Ain't Chicago: Race, Class, and Regional Identity in the Post-Soul South*. Chapel Hill, NC: University of North Carolina Press.

Roh, D. S., Huang, B., and Niu, G. A. 2015. Technologizing Orientalism: An Introduction, in D. S. Roh, B. Huang, and G. A. Niu (eds), *Techno-Orientalism: Imagining Asia in Speculative Fiction, History, and Media*. New Brunswick, NJ: Rutgers University Press, pp. 1–20.

Rose-Redwood, R. 2011. Rethinking the Agenda of Political Toponymy. *ACME: An International E-Journal for Critical Geographies* 10(1): 41–3.

Rose-Redwood, R., Alderman, D., and Azaryahu, M. 2010. Geographies of Toponymic Inscription: New Directions in Critical Place-name Studies. *Progress in Human Geography* 34(4): 453–70.

Roy, A. 2011. Slumdog Cities: Rethinking Subaltern Urbanism. *International Journal of Urban and Regional Research* 35(2): 223–38.

Rusk, D. 1993. *Cities without Suburbs*. Washington, DC: Woodrow Wilson Center Press.

Rutland, T. 2010. The Financialization of Urban Redevelopment. *Geography Compass* 4(8): 1167–78.

Ryberg-Webster, S. and Kinahan, K. 2017. Historic Preservation in Declining City Neighbourhoods: Analysing Rehabilitation Tax Credit Investments in Six US cities. *Urban Studies* 54(7): 1673–91.

Said, E. W. 1979. *Orientalism*. New York: Vintage Books.

Sassen, S. 2001 [1991]. *The Global City: New York, London, Tokyo*. Princeton, NJ: Princeton University Press.

Sauer, C. O. 1925. The Morphology of Landscape. *University of California Publications in Geography* 2(2): 19–54.

Schein, R. H. 1997. The Place of Landscape: A Conceptual

Framework for Interpreting an American Scene. *Annals of the Association of American Geographers* 87(4): 660–80.

Schneider, K. 2015. Long a College Town, Worcester Now Looks the College Part. *New York Times*, 6 January. Accessed online November 19, 2020 at https://www.nytimes.com/2015/01/07/realestate/commercial/long-a-college-town-worcester-now-looks-the-part.html?searchResultPosition=8/.

Schnell, I. and Haj-Yahya, N. 2014. Arab Integration in Jewish–Israeli Social Space: Does Commuting Make a Difference? *Urban Geography* 35(7): 1084–104.

Schoene, M. 2017. Urban Continent, Urban Activism? European Cities and Social Movement Activism. *Global Society* 31(3): 370–91.

Schön, D. A. 1980. Framing and Reframing the Problems of Cities, in D. Morley, S. Proudfoot, and T. Burns (eds), *Making Cities Work*. London: Croom Helm, and Boulder, CO: Westview Press, pp. 31–65.

Shaner, B. 2018. Time to Talk about Gentrification in Worcester. *Worcester Magazine*, 11 October. Accessed online June 22, 2022 at https://eu.worcestermag.com/story/news/2018/10/11/feature-time-to-talk-about-gentrification-in-worcester/9577761007/.

Short, J. R., Benton, L. M., Luce, W. B., and Walton, J. 1993. Reconstructing the Image of an Industrial City. *Annals of the Association of American Geographers* 83(2): 207–24.

Shtern, M. 2016. Urban Neoliberalism vs. Ethno-national Division: The Case of West Jerusalem's Shopping Malls. *Cities* 52: 132–9.

Simone A. 2009. *City Life from Jakarta to Dakar: Movements at the Crossroads*. London: Routledge.

Sites, W. 2012. "We Travel the Spaceways": Urban Utopianism and the Imagined Spaces of Black Experimental Music. *Urban Geography* 33(4): 566–92.

Smith, C. 2015. Art as a Diagnostic: Assessing Social and Political Transformation through Public Art in Cairo, Egypt. *Social & Cultural Geography* 16(1): 22–42.

Smith, M. E., Dennehy, T., Kamp-Whittaker, A., Stanley, B. W., Stark, B. L., and York, A. 2016. Conceptual

Approaches to Service Provision in Cities throughout History. *Urban Studies* 53(8): 1574–90.

Smith, M. P. 2001. *Transnational Urbanism: Locating Globalization*. Oxford: Blackwell.

Smith, N. 2005 (1996). *The New Urban Frontier: Gentrification and the Revanchist City*. Abingdon: Routledge.

Snow, D. A. and Benford, R. D. 1992. Master Frames and Cycles of Protest, in A. Morris and C. McClurg Mueller (eds), *Frontiers in Social Movement Theory*. New Haven, CT: Yale University Press, pp. 133–55.

Snow, D. A., Rochford Jr, E. B., Worden, S. K., and Benford, R. D. 1986. Frame Alignment Processes, Micromobilization, and Movement Participation. *American Sociological Review* 51(4): 464–81.

Soja, E. 2000. *Postmetropolis: Critical Studies of Cities and Regions*. Oxford and Malden, MA: Blackwell.

Solinger, D. J. 1999. *Contesting Citizenship in Urban China: Peasant Migrants, the State, and the Logic of the Market*. Berkeley and Los Angeles, CA: University of California Press.

Sorokin, P. A. 2002. Foreword, in F. Tönnies, *Community and Society: Gemeinschaft and Gesellschaft*, trans. and ed. C. P. Loomis. Minneola, NY: Dover Publications, pp. vii–viii.

State of Oregon. 2002. *Oregon Blue Book: Almanac and Fact Book*. Office of the Secretary of State. Accessed online June 21, 2022 at https://sos.oregon.gov/blue-book/Documents/elections/initiative.pdf.

Stephan, K. 2017. Portland Isn't Portlandia. It's a Capital of White Supremacy. *Washington Post*, 1 June. Accessed online May 5, 2022 at https://www.washingtonpost.com/opinions/the-hate-crime-in-super-progressive-portland-should-surprise-no-one/2017/06/01/d3b99782-46d8-11e7-a196-a1bb629f64cb_story.html/.

Su, F. and Tao, R. 2017. The China Model Withering? Institutional Roots of China's Local Developmentalism. *Urban Studies* 54(1): 230–50.

Sugrue, T. 2005 [1996]. *The Origins of the Urban Crisis:*

Race and Inequality in Postwar Detroit. Princeton, NJ: Princeton University Press.

Sun, W. 2014. *Subaltern China: Rural Migrants, Media, and Cultural Practices*. Lanham, MD: Rowman & Littlefield.

Suparak, A. 2021. *Virtually Asian*. Accessed online June 22, 2022 at https://vimeo.com/503907394/.

Surborg, B., VanWynsberghe, R., and Wyly, E. 2008. Mapping the Olympic Growth Machine. *City* 12: 341–55.

Swider, S. 2015. *Building China: Informal Work and the New Precariat*. Ithaca, NY: Cornell University Press.

Swyngedouw, E., Kaika, M., and Castro J. E. 2002. Urban Water: A Political-Ecology Perspective. *Built Environment* 28(2): 124–37.

Tao, R., Su, F., Liu, M., and Cao, G. 2010. Land Leasing and Local Public Finance in China's Regional Development: Evidence from Prefecture-level Cities. *Urban Studies* 47(10): 2217–236.

Tapp, R. and Kay, K. 2019. Fiscal Geographies: "Placing" Taxation in Urban Geography. *Urban Geography* 40(4): 573–81.

Teaford, J. 2006. *The Metropolitan Revolution*. New York: Columbia University Press.

Thoreau, H .D. 1854. *Walden Pond*. Boston: Ticknor and Fields.

Thrall, N. 2021. A Day in the Life of Abed Salama. *New York Review of Books*, 19 March. Accessed online April 29, 2021 at https://www.nybooks.com/daily/2021/03/19/a-day-in-the-life-of-abed-salama/.

Till, K. E. 2005. *The New Berlin: Memory, Politics, Place*. Minneapolis, MN: University of Minnesota Press.

Tönnies, F. 1955 [1887]. *Community and Association [Gemeinschaft and Gesellschaft]*. London: Routledge and Kegan Paul.

Tuan, Y. F. 1975. Place: An Experiential Perspective. *Geographical Review* 65(2): 151–65.

Uitermark, J., Nicholls, W., and Loopmans, M. 2012. Cities and Social Movements: Theorizing Beyond the Right to the City. *Environment and Planning A* 44(11): 2546–54.

Ueno, T. 1996. Japanimation and Techno-Orientalism. *Proceedings of the Seventh International Symposium on Electronic Art*, Rotterdam, Netherlands, 16–20 September 1996. ISEA96 Foundation. pp. 94–6. Accessed online June 10, 2022 at https://isea-archives.siggraph.org/symposium/isea96-seventh-international-symposium-on-electronic-art/.

Valentine, G. 2013. Living with Difference: Proximity and Encounter in Urban Life. *Geography* 98(1): 4–9.

Valverde, M. 2012. *Everyday Law on the Street: City Governance in an Age of Diversity*. Chicago, IL: University of Chicago Press.

Verma, P. and Raghubanshi, A. S. 2018. Urban Sustainability Indicators: Challenges and Opportunities. *Ecological Indicators* 93: 282–91.

Wang, Z. 2020. Beyond Displacement – Exploring the Variegated Social Impacts of Urban Redevelopment. *Urban Geography* 41(5): 703–12.

Ward, C. and Swyngedouw, E. 2018. Neoliberalisation from the Ground Up: Insurgent Capital, Regional Struggle, and the Assetisation of Land. *Antipode* 50(4): 1077–97.

Ward, K. 2006. "Policies in Motion," Urban Management and State Restructuring: The Trans-local Expansion of Business Improvement Districts. *International Journal of Urban and Regional Research* 30: 54–75.

Ward, S. 1998. *Selling Places: The Marketing and Promotion of Towns and Cities 1850–2000*. London: Routledge.

Weber, R. 2010. Selling City Futures: The Financialization of Urban Redevelopment Policy. *Economic Geography* 86(3): 251–74.

Weber, R. and O'Neill-Kohl, S. 2013. The Historical Roots of Tax Increment Financing, or How Real Estate Consultants Kept Urban Renewal Alive. *Economic Development Quarterly* 27(3): 193–207.

Wilder, C. S. 2014. *Ebony and Ivy: Race, Slavery, and the Troubled History of America's Universities*. New York: Bloomsbury Press.

Wilson, E. 1991. *The Sphinx in the City: Urban Life, the Control of Disorder, and Women*. Berkeley and Los Angeles: University of California Press.

Wilson, K. 2004. *Livable Modernism: Interior Decorating and Design during the Great Depression.* New Haven, CT: Yale University Press.

Wirth, L. 1938. Urbanism as a Way of Life. *American Journal of Sociology* 44: 1–24.

Wolsink, M, 2020. Framing in Renewable Energy Policies: A Glossary. *Energies* 13(11): 2871.

Wright, W. J. and Herman, C. K. 2018. No "Blank Canvas": Public Art and Gentrification in Houston's Third Ward. *City & Society* 30(1): 89–116.

Wu, F. 1999. The "Game" of Landed-Property Production and Capital Circulation in China's Transitional Economy, with Reference to Shanghai. *Environment and Planning A* 31(10): 1757–71.

Wu, F. 2021. The Long Shadow of the State: Financializing the Chinese City. *Urban Geography* DOI: 10.1080/02723638.2021.1959779.

Wu, F., Xu, J., and Yeh, A. G.-O. 2006. *Urban Development in Post-reform China: State, Market, and Space.* London: Routledge.

Xu, J., Yeh, A. G. O., and Wu, F. 2009. Land Commodification: New Land Development and Politics in China since the Late 1990s. *International Journal of Urban and Regional Research* 33(4): 890–913.

Yang, C. 2017. The Rise of Strategic Partner Firms and Reconfiguration of Personal Computer Production Networks in China: Insights from the Emerging Laptop Cluster in Chongqing. *Geoforum* 84: 21–31.

Ye, J. and Pan, L. 2011. Differentiated Childhoods: Impacts of Rural Labor Migration on Left-behind Children in China. *Journal of Peasant Studies* 38(2): 355–77.

Yiftachel, O. 2016. The Aleph – Jerusalem as Critical Learning. *City* 20(3): 483–94.

Young, I. M. 1990. *Justice and the Politics of Difference.* Princeton, NJ: Princeton University Press.

Zelinsky, W. 1997. Along the Frontiers of Name Geography. *Professional Geographer* 49(4): 65–6.

Zelinsky, W. 2002. Slouching Toward a Theory of Names: A Tentative Taxonomic Fix. *Names* 50: 243–62.

Zhang, L. 2001. *Strangers in the City: Reconfigurations of Space, Power, and Social Networks within China's Floating Population*. Stanford, CA: Stanford University Press.

Zhang, L. 2012. Economic Migration and Urban Citizenship in China: The Role of Points Systems. *Population and Development Review* 38(3): 503–33.

Zhang, L. and Tao, L. 2012. Barriers to the Acquisition of Urban Hukou in Chinese Cities. *Environment and Planning A* 44(12): 2883–900.

Zhang, T. 2002. Urban Development and a Socialist Pro-growth Coalition in Shanghai. *Urban Affairs Review* 37(4): 475–99.

Zhang, Y. 2018. Grabbing Land for Equitable Development? Reengineering Land Dispossession through Securitising Land Development Rights in Chongqing. *Antipode* 50(4): 1120–40.

Zhang, Y. 2020. The Chongqing Model One Decade On. *Made in China Journal* 5(3): 31–9.

Zhu, J. 2005. A Transitional Institution for the Emerging Land Market in Urban China. *Urban Studies* 42(8): 1369–90.

Zimmermann, K., Galland, D., and Harrison J. (eds). 2020. *Metropolitan Regions, Planning and Governance*. Switzerland: Springer.

Zukin, S. 1989. *Loft Living: Culture and Capital in Urban Change*. New Brunswick, NJ: Rutgers University Press.

Zukin, S. 1995. *The Cultures of Cities*. Cambridge, MA: Blackwell.

Zukin, S. 2009. Changing Landscapes of Power: Opulence and the Urge for Authenticity. *International Journal of Urban and Regional Research* 33(2): 543–53.

Index

Page numbers in **bold** refer to boxes in the text. Page numbers in *italic* refer to photographs.

Aberdeen, United Kingdom 17
activism **11**, 65, 69–73, 107, 179
  Portland 112, 120–1
  protests
    Jerusalem 164, 174
    and racial issues 23, 98, 120–1
  Worcester 80, 83–4
actors, rational **15**
advocacy, community 57
Afghani people 64
agency, human **15**–17
  and place framing 14–16, 185–6
agglomeration, urban 59–61
Agnew, John 7–9
Albina, Portland, Oregon 110
Alderman, D. H. and Inwood, J. **155**
Al-Haj, M. **65**
analysis, urban 13–17, 182
Animal and Plant Health Inspection Service (APHIS) 86, 87–8, 90
anonymity 62–3

Anoplophora glabripennis (longhorned beetle) 85–8, 94
Arab Israelis **157**, 194
archeology, Old City, Jerusalem 168–9
architecture 127
  London 36, **39**, 41, **131**
  Old City, Jerusalem 159, 168
Armenian people 64, 161, 193n9
Arreola, Daniel 43
art/artists 5, 43–5, 65, 67
atomic power **39**
automobility **39**, **101**
Avila, Eric **121**

Baltimore, Maryland, United States 16
banks/banking 19, 25, 60, 193n5
Barbican Center 41–2
Barbican Estate, London 27, 32–42
  and cosmopolitanism 42, 44

Barnes, Trevor and Sheppard, Eric 188–9
baseball 80–2, 83–4
basketball 187
Baudrillard, Jean **139**
Beaverton, Oregon 187
beetles, longhorned 85–8, 94
Beijing, China 140
Belfast, United Kingdom 154, 171
beliefs/intentions 4
belonging **129**
Bilbao, Spain **131**
biophysical processes **91**
Birmingham, United Kingdom 17
birth rates 49
Black Lives Matter *118*
Black people, United States **15**, 67, **117**, **119–21**, 181
  Portland 106–12, 114–16, 120
Bledsoe, Adam and Wright, Willie **119**
Bo Xilai 146–8, 150
Boston, United States 72
Boston Globe 78
boundaries **43**, 187–8
Bourdieu, Pierre **15**
branding 78, **129–33**
  rebranding 78, **131**
Brisbane, Australia 2
Britishness 31
Broadbent, Jeff **55–7**
builders, local 54–6
built environment, futuristic **137–41**
  Chongqing 24, 126–38, 148, 184
bundles
  and place framing 6, **9**, 16–17, 180
  of space–time trajectories 4–5, 6

buses 20
  see also transport, public
business leaders 57
businesses, small **55**

canal district, Worcester 80–2
capitalism **15–17, 29–31**, 189
  City of London 22, 27–31, 44–5, 46
  and place framing 10, 44–5
car use **39, 101**
Carson, Rachel **107**
Castells, Manuel **5**
Çatal Hüyük, Turkey **5**
celebrations 4
  religious 162–3, 166, 195n2, 195n3
Census, Iranian 49, 60
Census, US 72, 73
central business districts (CBDs) 128, 133
chadors 63, 193n6
change, social **5, 11**, 191
Chauvin, Derek 120
Chicago, United States **5**, 87
Chicago School **115–17**
China
  citizenship 146, **147**
  economic development 124–5
  economic development/growth 126, 141, **143–5, 144**
  governance 127, **149–51, 150**
  household income 140, 141, **147**
  household registration system (*hukou*) 140–1, 142–4, **143–7**
  industrialization 124, 125, **143**
  inequality 138–48, 144–6, **145–7, 150**

infrastructure 124, **151**
  public transport 127, 132, 136, 138
 land ownership 146, **149–51**
 land use **57**, **149**
 migrants 140–2, **143–7**, 144–6
  and urban/municipal services 141–2, **145–7**
 New Areas 124–5, 141
 public rental housing 141–2, 144, 148, 150
 rural land 142–4, 146, **149**, 150
 temporary residence permits 140, **147**
 urban development **57**, **137**, **143–5**, **149–51**
  Chongqing 24, 124–5, 126, 135, 138–48
choice, personal 16, 158, 185
Chongqing, China 23–4, 123–52
 central business districts 128, 133
 futuristic built environment 24, 126–38, 148, 184
  Asian futurism 135, 136, 148
  tower blocks 128–32, 138
  urban/vertical landscapes 127–32, 133, 135
  weirdness 132–3, 136–8
 household income 140, 141
 industrialization 124, 125
 inequality 138–48, 150
 infrastructure 124
  public transport 127, 132, 136, 138
 migrants 140–2, 144–6
 municipal government 125, 126, 184
  and *hukou* 140, 141–4, 146, 150
  and tourism 136, 138
 public rental housing 141–2, 148, 150
 residents 128, 136–48
 rural land/farming 142, 146, 150
 social media 24, 127, 132–3, 148
 tourism 24, 132, 136, 138, 184
 urban development 24, 135, 138–48
  and economic development 124–5, 126
Chongqing County Land Exchange 142
"Chongqing Model" 146
Christianity, Jerusalem
 and residents 153, 171
 and tourism 159, 160, 161
Church of the Holy Sepulcher, Jerusalem 159, 162
cities, definition 1
cities, global 35
citizenship
 China 146, **147**
 Israel 154, **157**, 171, 172, **194n1**
city leaders, Worcester 74, 80, 83
City of London Corporation 26, 34–6, 42
City of London, United Kingdom 25–47
 capitalism 22, 27–31, 44–5, 46
 and household income 36
 inequality 39, 45
 international finance 30–1, 35, 46, 50
 international trade 26, 28–30
 internationalism 44, 46, 182
 lifestyle 46–7, 177–8

City of London, United
Kingdom (*cont.*)
  modernist urbanism 22, 27,
    36–42, 45, 46–7
  cosmopolitanism 42, 44
  political economics 22, 25,
    26, 27–31, 44–5, 46,
    182
  residents 27, 31, 44
    social class 32–4, 36, 41,
      45–6
    wealth 27–8, 32, 42, 46–7,
      177–8, 182
CitySquare, Worcester 75, 76
climate, Tehran 64–7
clothing/dress
  Old City, Jerusalem 159,
    160, 162
  Tehran 63, 66, 193n6,
    193n7
colonialism 30, 50
  Portland 23, 98, 106–12,
    114–16
color, use of 37
Commission for Development
  and Reform, China 144
Common Council, City of
  London 36
community action 67, 94–5
community support/networks
  5, 15, 67, 77, 146, 177,
  189
commuting 32–4, 49, 102
  Worcester 72, 74
"comparative advantage" 50
competitiveness **129**, **131**
completeness *see*
  incompleteness
computers 125
conceptual framing **11–13**
  see also place framing
conflict 179
  Jerusalem 154–5, 170–1,
    174

connectivity 2, **5**, **7–9**, 94–5,
  177, 189
conservationism **107**
conservatism 99, **107**
construction workers, China
  145
consumption 45
  Old City, Jerusalem 160,
    162
  Worcester 83, 95–6
coordination, political **57**
corruption 12, 56
cosmopolitanism
  City of London 42, 44, 46
  Tehran 23, 49, 60–7, 68, 70,
    182, 183
counterculture 99
counties, China 123–4
creativity **5**, **65**, 78
Cresswell, Tim 7
Cronon, William **89**
cultural diversity 60, **65**, 68–9
cultural geography, urban
  43–5
culture 1, 75, 99
  Asian futurism **137–41**
  and identity **43–5**, 166
  religious 63–4, 66, 166
  Tehran 63–4, 66, 68
cyberpunkism, Chongqing 24,
  126–38, 148

Dallas, Texas 105
Damascus Gate, Jerusalem
  164–6
Darwinism **115**
Declaration of Independence,
  Israeli 194n1
demonyms (names for people),
  Israel 154, **155–7**,
  194n1
density 2, **3**, 6–8, **101–3**, 177
  Portland 100, 102, 103–5,
    112

Department of Conservation and Recreation, Massachusetts 88, 90
design, urban 37–9
Detroit, Michigan 105, **117**, **129**
developers 54–6, **79**, **151**
 Worcester 74, 76–8, 83
development, economic
 China 124–5, 126, **143**–5
 Tehran 49–52, 70–1
 banks/banking 60, 193n5
development, municipal **77–81**, **183**
development, sustainable 97, 98, 102, 105–6, 112
development, top-down 51
development, urban 32–6, 100
 China 57, 137, **143**–5, **149**–51
 Chongqing 24, 135, 138–48
 redevelopment 32, 76, **77–81**, 183
 and tourism **81**, 87
 urban development strategies 77, 97, **129**
developmentalism, Islamic 48–71
difference, social 12, **43**–5, **111**–13
 Tehran 62, 63, 183
disinvestment 65, **79**, 110, **113**
disorder 12, 43
displaced people
 Chongqing 124, 146
 Jerusalem 163, 171, 172
displacement 178
district improvement financing/tax-increment financing 76, **77**–9, 82
districts, China 123–4
diversity, cultural 60, **65**, 68–9

diversity, religious 64, 68, 159–60, 193n9
diversity, social 64, **65**–7, 70
dockyards, Portland, Oregon 99
Dome of the Rock mosque, Jerusalem 159, 164
Druze people 154, 194n1
Du Bois, W. E. B. **117**

Eames chairs 41, 192n5
ecomodernism **107–9**
 Portland 23, 98, 99–106, 112–14, 121
economic development/growth 10, **33–5**
 China 124–5, 126, 141, **143**, 144
 Tehran 22–3, 49–61, 68, 70–1
economic processes 1, 10, **33–5**, 59, 185, 189
economic structure 14–16, **15–17**, 185–6
economic systems 15, 19, 29, 186, 189
economics, political 15, 16, **29–31**
 City of London 22, 25, 26, 27–31, 44–5, 46, 182
 Worcester 95–6
 economic restructuring 23, 72–96, 183
economy, global 35, 48
educational opportunities, Worcester 74, 78
efficiency 39
elections 44
electricity supplies, Tehran 58
elevators 39, 128
elites, urban 12, 68, 80, 85
emplacement 178
employment opportunities 60, 83–4, 177

engaged pluralism 188–9
Engels, Friedrich **29**
entertainment 54, 74, 83–4
entrepreneurialism **33**, **129–31**
environmental issues **81**, 177
  environmental sustainability 90
    Portland 97, 100–2, 105–6, 112
    New York City 86, **89**
    Portland 97–102, 105–6, 112, 183–4
    Worcester 23, 73, 85–93, 183
environmental policy, Worcester 85–90, 93
environmentalism, traditional **107–9**
environments, urban **89–91**
Esfahani, Azadeh Hadizadeh 48–71, 182
essentialism 135
Europe 37, 42, 176
experience, geographical 7, 119
expression, cultural **43–5**, 67, 68

factory workers, China 145
Fair Housing Act, United States 113
families, nuclear **65**, 99
Fang, C. and Yu, D. **59**
farming
  Chongqing 142, 146
  Portland 99, 102, 105
feminist thinking 10, **15**, 180
finance, international 30–1, 46, 50
finance, land-based **151**
finance strategies **77–81**
financialization of urban policy **33–5**
Floyd, George 98, 116–22, 184

Foreigners' Street, Chongqing 127
forests **69**, **91**
  New England 86–90, 94
  Portland 102, 105
  Worcester 73, 86, 88, 90–5
Foxconn 125
fragmentation of urban processes 179–80
framing, definition of **11**
  see also place framing
France **101**
freedom, social 62–3, **65–7**, 68
functionalism **37–9**
futurism, Asian **137–41**
  Chongqing 135, 136, 148

Gandy, Matthew **89**
gas, Tehran 58, 60
gender issues 10, 180
gentrification 14, **31**, **67**, 82
geography 6, 7, **43–5**, 64–7, **119–21**
Gibson-Graham, J. K. **15**
Giddens, Anthony **15**
Global North 50, **107**
Global South 22, 48, 176
governance 3, 25–6, **33–5**, **129**
  China 127, **149–51**, **150**
  Tehran 48, 49
government, municipal 56
  Chongqing 125, 126, 184
  and *hukou* 140, 141–4, 146, 150
  and tourism 136, 138
government funding, Worcester 76, 82
government officials 54, 73–4
graffiti **43–5**, 127, 173
grassroots involvement, Worcester 94–5
Great Depression 111, 114

Index

Great Western Development Strategy, China 124
Greater London Authority 25, 45
Greater London Plan 34
Greening the Gateway Cities program, Massachusetts 93
Guggenheim Museum, Bilbao **131**

Haganah, Jewish (Zionist) paramilitary organization 159
Hanover Theater for the Performing Arts, Worcester 75
Harvey, David **129**, 144
Hewlett Packard 125
Hong Kong 132
house prices, Tehran 52–4
household registration system (*hukou*); see hukou (household registration system), China
housing, luxury, Worcester 78, 82
housing, public rental, China 141–2, 144, 148, 150
housing, social, City of London 27
housing programs, federal **111**–13
Hualongqiao station 136, 138
Huangjueping Main Street, Chongqing 127
hukou (household registration system), China 140–4, **143**–7, 150
Hunter, Marcus Anthony **15**, **121**, 180–1

identity, collective 7, **11**, 67
and branding **129**, **131**–3
Portland 99, 102, 121
religious **43**–5, **155**–7
Jerusalem 24, 166, 184
Worcester 23, 78, 82, 94–5
identity, individual 7
income, household
and inequality 60, **81**, **111**
City of London 39, 45
lower-income households 36, 60
China 140, 141, **147**
Worcester 93, 94
incompleteness 95, 121, 158, 171, 174
and place framing 10, 16, 181, 186–8
Indigenous people 6, 98, 114
"industrial linkage and spillover" model **149**
industrialization 28, 73
China 124, 125, **143**
Tehran 51, 52
inequality **29**, **61**, 93, **133**, 173
China 138–48, **145**–7, 150
City of London 39, 45
and household income 39, 45, 60, **81**, **111**
Iran 58–60, 69
and place framing 180, 185
infrastructure **59**, 183
China **151**
Chongqing 124, 127, 132, 136, 138
Worcester 75, 96, 183
see also transport, public
interest rates, Iran 193n5
international relations, Iran 52
International Trade and Commerce Center, Chongqing 136
internationalism, City of London 44, 46, 182
interrogating cities 178–84

investment 54–6, **65–7**
  disinvestment 65, **79**, 110, **113**
  private sector 76–8, **79**, 80
  speculative **29–31**, **149**
  Worcester 73, 76–8, 80
Iran 51, 52, 58–60
  Iranian revolution 52, 62
  oil production 48, 52, 54
  see also Islam; Tehran, Iran
Islam 48–71
  Jerusalem 153, 159, 166, 171
  Tehran 23, 48, 62, 68, 70, 182–3, 193n5
  religious culture 63–4, 66
Israel 153, 173–4
  citizenship 154, **157**, 171, 172, **194n1**
  names for people and places 154, **155–7**, 194n1
  nationalism 159, 163, 170, 184
  residents
    Jewish citizens of Israel 153, 154, 166, 194n1
    Palestinian citizens of Israel 154, 160–1, 172
  see also Jerusalem, Israel; Judaism, Israel; Old City, Jerusalem
Israeli people 154, 163, 194n1
  see also Palestine/Palestinian people
Israeli/Palestinian conflict 170–1

Jaffa Gate, Jerusalem 160, 161
Japan **137**, **139**
Jerusalem, Israel 24, 153–75
  activism 164, 174
  conflict 154–5, 170–1, 174
  demonyms (names for people) 154, **155–7**, 194n1
  displaced people 163, 171, 172
  religious culture 24, 153–75
    Christianity 153, 159, 160, 161, 171
    Islam 153, 159, 166, 171
    Old City, Jerusalem 159–60, 162–3, 166, 174, 184
  religious identity 24, 166, 184
  residents 153, 171–3
  segregation, social/political/religious 164, 172–3
  tourism 24, 156–69, 184
  see also Old City, Jerusalem
Jewish people, Israel
  Jewish citizens of Israel (residents) 153, 154, 166, 194n1
  Old City, Jerusalem (tourists) 159, 160
John, King 26
Judaism, Israel 153
  Old City, Jerusalem 159, 160, 161–2, 163–4, 169, 170–1
  and collective identity 24, 166, 184
  Orthodox 159, 160, 163, 164

Kaiser, Henry 109
King, Martin Luther **155**
Kotel, Jerusalem (Western Wall) 159, 163–4, *165*, 168
Ku Klux Klan 109
Kurtz, Hilda **11**

land, rural 89, 103
  rural land exchange program, China 142–4, 146, **149**, 150
land appreciation, China 142
land banks, China 142

land development 55, 61
land markets **149–51**
land ownership 146, **149–51**, 171–2
land use 56, **61**, 103, **111**, 179
  China 57, **149**
land value 29, 77–9, **111–13**, **151**
  Tehran 56, 57–8, 69
"land-based finance" **151**
landscapes, cultural **167–9**
landscapes, urban 13–14, **89**, 177
  Chongqing 127–32, 135
  Portland **105**, 110
landscapes, vertical 127–32, 133
landscapes, wilderness **107**
Lefebvre, Henri 5, **45**, 46, **65**, 179–80
legislation 100, **111**, 112, **113**
Liangjiang New Area, Chongqing 124, 125, 141
life, everyday 85, 138, **167–9**, 174
lifestyle
  City of London 46–7, 177–8
  Tehran 64, 69, 70, 177–8, 183
Liziba station, Chongqing 132, 134, 136, 138
Lloyd's of London 28–30
London, United Kingdom 22, 25–47, 52, 192n2
  architecture 36, **39**, 41, **131**
  see also City of London, United Kingdom
Lord Mayor of London 25–6, 36
Los Angeles, California 50, **129**
Louisiana, United States 11

McGovern, Jim 88
McKittrick, Katherine 119

Magna Carta 26
malls, downtown/shopping 74–5, 84, 160–1
Mamilla Mall, Jerusalem 160
Manila, Philippines 2
mantoos 63, 193n7
maple sugar industry, Canada 88
market economy **143–5**, 186
marriage, Tehran 64, 193n8
Martin, Deborah 11, **13**, **167**
Marxism 10
Massachusetts College of Pharmacy and Health Science University 74
Massey, Doreen 4, 7, **9**, 179–80, 190–1
master plans, Tehran 52
Mayor of London 25
meaning **115**, 166
  and place framing 7–**9**, 10, 11, 12, 21
media industry 50, 55, 78, **129**, 148
  social media 24, 127, **129**, 132–3, 148
memorialization **167**
Metro, Oregon 100
migrants 12, **43**, 82–3
  China 140–2, **143–7**, 144–6
  and urban/municipal services 141–2, **145–7**
  and racial issues **115**, **121**
  Tehran 49, 52, 60, 64, 68
Ministry for the Development of Urban and Rural Housing, China 144
Ministry of Finance, China 144
modernism
  ecomodernism **107–9**
  Portland 23, 98, 99–106
  modernist urbanism **37–9**
  City of London 22, 27, 36–42, 44, 45, 46–7

Mohammad Reza Shah Pahlavi 52
Molotch, Harvey 55
monarchy, British 26
monopolies, City of London 30
monuments **167**
moralizing 12
mortgages **111–13**
"Mountain City" 127, **129**
　see also Chongqing, China
Muir, John **107**
multiculturalism, Tehran 68
music **45**

names for people and places, Israel 154, **155–7**, 194n1
National Mall, Washington, DC. 4
nationalism 182–3
　Israel 159, 163, 170, 184
native Americans 98, 114
nature see environmental issues
neighborhoods, residential
　see residential areas/neighborhoods
New Areas, China 124–5, 141
New England forests 86–90, 94
New Gate, Jerusalem 160–1
New York City, United States 2, **35**, 45, **52**, **129**
　environmental issues 86, **89**
New York Times 78
New York Times Magazine 82
New-type Urbanization Plan, China 144
Nike 187
non-residents, City of London 44
Norman Conquest 26, 192n1
Noxolo, Pat 190

oil production, Iran 48, 52, 54
Oita Prefecture, Japan 57

Oklahoma City, United States 2
Old City, Jerusalem 24, 158–69
　architecture 159, 168
　consumption 160, 162
　Palestine/Palestinian people 160–1, 169, 184
　place framing 170–3, 174–5
　power 168–9, 170–3
　religious culture 174, 184
　　clothing/dress 159, 160, 162
　　religious celebrations 162–3, 166
　religious diversity 159–60
　residents 164, 170–1, 195n5
　and tourism 168, 169
　security 163, 166
　tourism
　　Christianity 159, 160, 161
　　Islam 159, 166
　　Jewish people 24, 156–9, 184
　　and residents 168, 169
　see also Israel; Jerusalem, Israel
Old City walls, Jerusalem 168
Olmsted, Frederick Law **89**
Olympic Games bids 87
Open Up the West program, China 124, 128
Oregon, United States 99
　legislation (SB100) 100, 112
　racial issues 106–9, 110, 114, 116
　state constitution 108, 116
　see also Portland, Oregon, United States
orientalism **137–41**

Pacific Northwest forests 102
Palestine/Palestinian people 153

Old City, Jerusalem 160–1, 164, 169, 184
  residents 155–7, 164, 171–4, 195n5
  Palestinian citizens of Israel 154, 160–1, 172
Palestinian Authority 172
Paris, France 45
parking areas 101
parks, urban 89
paternalism 50
Pawtucket Red Sox 80–2, 84
people in cities 19–20
Philadelphia, United States 117
place, relational 82, 119, 187, 190
place conflict 179
  Jerusalem 154–5, 170–1, 174
place framing 6–17, 21–2, 24, 176–91
  and bundles 6, 16–17, 180
  and capitalism 10, 44–5
  and conceptual framing 11–13
  and feminist thinking 10, 15, 180
  and human agency 14–16, 185–6
  and incompleteness 10, 16, 181, 186–8
  and inequality 180, 185
  and meaning 7–9, 10, 11, 12, 21
  and personal choice 16, 185
  and plurality 17, 112, 179–80, 185, 188–9
  and politics of place 12–14, 112, 178, 179, 181, 190–1
  and power 22, 170–3, 191
  and social class 42, 180
  and urban analysis 13–17, 182

  see also Chongqing, China; Jerusalem, Israel; London, United Kingdom; Portland, Oregon, United States; Tehran, Iran; Worcester, Massachusetts, United States
place theory 7–9
placemaking 2–5, 21, 174, 177, 178–9, 180–1, 190, 191
  and racial issues 119, 181
places, whole 178
planning, urban 37, 84, 87, 111–13, 149
plurality 179–80
  radical 17, 112, 185, 188–9
Polar Park, Worcester 82
policing, Portland, Oregon 98, 120
policy mobilities 87
political economics *see* economics, political
political participation, Tehran 69–70
political systems 3, 46, 91
politics 1, 29–31, 81
  left-wing, Portland 23, 97–8, 99–100, 102, 110–12, 121
  politics of place 12–14, 112, 178, 179, 181, 190–1
  Worcester 23, 73–4
population numbers 105, 123, 176
  Tehran 49, 52, 58, 60, 192n1
  Worcester 72, 73
Portland, Oregon, United States 23, 97–122, 187–8
  activism 112, 120–1
  colonialism 23, 98, 106–12, 114–16

Portland, Oregon, United States (*cont.*)
  density 100, 102, 103–5, 112
  ecomodernism 23, 98, 99–106, 112–14, 121
  environmental issues 97–102, 183–4
    and sustainability 97, 98, 100–2, 105–6, 112
  environmental urbanism 99–103, 105–6, 183
  farming 99, 102, 105
  forests 102, 105
  identity, collective 99, 102, 121
  Indigenous people 98, 114
  left-wing politics 23, 97–8, 99–100, 102, 110–12, 121
  policing 98, 120
  public transport 97, 100–2, *104*, 110
  racial issues 23, 98, 106–12, 114–22, 183–4
    and Black people 106–12, 114–16, 120
    and protests 23, 98, 120–1
    and racial erasure 106–12, 114, 116
    and white people 98, 106–12, 114, 121
  residents 116, 120, 121
  slave trade 106, 108
  sprawl, urban 97, 99
  suburbs 99–100, 102
  urban growth boundaries 97, 100, 103
  urban landscapes **105**, 110
  weirdness 102, 105
  World War II 99, 109, 114
  young people 102, 112, 121
Portland Police Bureau 120
postcard campaign, Worcester 80, 83–4

Postcolonial Urbanisms project **121**
poverty 45–6, **81**, 93
  poor people 14, **39**, 46, 69, **81**
power **167–9**, 174
  Old City, Jerusalem 168–9, 170–3
  and place framing 22, 170–3, 191
power, community **67**
power, western **141**
private sector investment 76–8, **79**, 80
privatization, Tehran 54
privilege 158
property ownership 56, **79**
property taxes **31**, **33**, 76, 77–81
propinquity 2, 6–8, 20, 177
protests 69–73
  Jerusalem 164, 174
  and racial issues 23, 98, *118*, 120–1

quality standards 36–41, **39**

racial issues 50, **115–17**
  and Black people **15**, **67**, 106–12, 114–16, **117**, **119–21**, 120, 181
  and migrants **115**, **121**
  and placemaking **119**, 181
  Portland 23, 98, 106–12, 114–22, 183–4
    and Black people 106–12, 114–16, 120
    Oregon 106–9, 110, 114, 116
    and protests 23, 98, 120–1
    and racial erasure 106–12, 114, 116
    and white people 98, 106–12, 114, 121

and racial segregation 109,
  111–13, 117
and residents 115–17, 116,
  120, 121
and United States 111–13,
  181
and white people 98,
  106–12, 113, 114, 117,
  119, 121
racism, environmental 91
racism, exclusionary 97–102
  see also Portland, Oregon,
  United States
Ramadan, Jerusalem 166
real estate 55, 65
  Tehran 54, 56, 68, 70
  Worcester 73, 75
rebuilding, Tehran 56, 193n4
reconstruction, postwar 37
Red Square, Moscow 4
redevelopment 32, 76, 77–81,
  183
redlining 111–13
reductionism 135–6
regeneration, urban 74–85,
  131
"re-imagineering" 131
relational place 82, 119, 187,
  190
religious culture
  celebrations 162–3, 166
  Islamic 63–4, 66
  Jerusalem 24, 153–75
    Old City, Jerusalem
    159–60, 162–3, 166,
    174, 184
    religious identity 24, 166,
    184
  religious diversity 64, 68,
    159–60, 193n9
  Tehran 64, 68, 193n9
  see also Christianity,
    Jerusalem; Islam;
    Judaism, Israel

relocation of companies 75, 125
residential areas/neighborhoods
  23, 32, 52, 73, 111,
  115–17, 128–32
residents 55, 67
  and branding 129, 133
  Chongqing 128, 138–48
  City of London 27, 31,
    32–4, 44
    social class 32–4, 36, 41,
    45–6
  Israel
    Jewish citizens of Israel
    153, 154, 166, 194n1
    Palestinian citizens of
    Israel 154, 160–1, 172
  Jerusalem 171–3
    Christianity 153, 171
    Islam 153, 171
  Old City, Jerusalem 164,
    170–1, 195n5
    and tourism 168, 169
  Palestine/Palestinian people
    155–7, 164, 171–4,
    195n5
  Palestinian citizens of
    Israel 154, 160–1, 172
  Portland 116, 120, 121
  and racial issues 115–17,
    116, 120, 121
  and social class 32–4, 36,
    41, 82, 93
  Tehran 56, 61–2, 69–71
  Worcester 80, 82, 83–4, 85,
    87–93, 94–5, 96
resources, Tehran 57–61
restructuring, economic,
  Worcester 23, 72–96,
  183
revolution, Iran 52, 62
Reza Khan 51
rivers, Tehran 58, 67
Robinson, Zandria 15, 121,
  180–1

Roman Gate, Jerusalem 166
Romans 26
rural land 89, 103
   rural land exchange program, China 142–4, 146, **149**, 150

Said, Edward **137**
Sassen, Saskia **35**
Sauer, Carl **167**
SB100, Oregon 100, 112
Schneider, K. 78
science fiction 133, 135
security **111**, 163, 166
segregation, gender 164
segregation, racial 109, **111–13, 117**
segregation, social and political 172–3
serendipitous engagement *see* propinquity
service workers, China **145**
services, urban/municipal 76, 77, **79**, 110
   and migrants, China 141–2, **145–7**
settlements, informal, Tehran 52
Seventh Ward, Philadelphia **117**
Shavuot festival 163, 195n2
Sheikh Jarrah, Jerusalem 171–2
shipbuilding, Portland, Oregon 109
Sichuan Province, China 123
Sinofuturism **137–41**
Six-Day War 163, 166, 171
slave trade 30, **121, 167**
   Portland 106, 108
slums 34, 110, 135, 176
Smith, N. **65**
Snow, D. A. **11**
social class
   and place framing 42, 180
   and residents
      City of London 32–4, 36, 41, 45–6
      Worcester 82, 93
social diversity 64, **65**–7, 70
social media **129**
   Chongqing 24, 127, 132–3, 148
social mobility, China **145–7**
social movements 6, **11**
   see also activism
social policy, Chongqing 140–1
social relations 60, 64, **65**
social services 120, 142, 193n4
Soja, Ed **5**
souvenir shops, Jerusalem 162
space–time trajectories 4–5, 6, 91
spatial form **59**
speculation **29–31**, 133, 149
sport 80–2, **83–4**, **87**
sprawl, urban 97, 99, **101–3**
St Vincent's Hospital, Worcester 74
state constitution, Oregon 108, 116
"state-subsidized workers' dormitories" 148
streams 8
street signs, Jerusalem 170
streets, city 18–19
subcommunities **15**
subsidies 58, **79, 81**
suburbs 32–4, **101**
   Portland 99–100, 102
   United States 37, **113**
Sukkot festival 163, 195n3
Suparak, Astria 135
superiority, western **137–9**
supremacy, white *see* racial issues; white people
sustainability, environmental **91**

Portland 97, 100–2, 105–6, 112
symbolism **155**, **167**, 170
Tehran 58, 69

tax incentives 33, **79–81**
tax revenues **149–51**
property taxes 31, 33, 76, **77–81**
Tehran 56, 193n4
Worcester 75–6, 80, 95–6
tax-increment financing/district improvement financing 76, **77–9**, 82
technologies, new **39**, 107–9, 125
Asian futurism **137**, 139
technologism 37
Tehran, Iran 22–3, 48–71, 182–3
Armenian people 64, 193n9
cosmopolitanism 23, 49, 60–7, 68, 70, 182, 183
diversity, cultural 60, 63–4, **65**, 66, 68–9
diversity, religious 64, 68, 193n9
diversity, social 64, 70
economic development/growth 22–3, 49–61, 68, 70–1
banks/banking 60, 193n5
governance 48, 49
industrialization 51, 52
inequality 58, 69
Islam 23, 48, 62, 68, 70, 182–3, 193n5
women/women's clothing 63–4, 66, 193n6, 193n7
land value 56, 57–8, 69
lifestyle 64, 69, 70, 177–8, 183
marriage 64, 193n8
migrants 49, 52, 60, 64, 68

population numbers 49, 52, 58, 60, 192n1
real estate 54, 56, 68, 70
rebuilding 56, 193n4
residents 56, 61–2, 69–71
rivers 58, 67
social difference 62, 63, 183
social freedom 62–3, 68
symbolism 58, 69
tax revenues 56, 193n4
universities 52, 68
urban growth coalition 56, 68, 70
utilities 58, 69
wealth 54, 60, 177–8
women's clothing 63, 66, 193n6, 193n7
young people 60, 68
Temples, Jerusalem 162, 164, 168
temporary residence permits, China 140, **147**
tensions, social, political and religious 68, 166, 189
Thoreau, Henry David **107**
Three Gorges Dam, China 124
Tokyo, Japan **35**, 52
toponyms (names for places), Israel 154, **155–7**, 194n1
Toronto, Canada 87
tourism 88
Chongqing 24, 132, 136, 138, 184
Jerusalem 24, 156–69, 184
and Christianity 159, 160, 161
and Islam 159, 166
and Jewish people 159, 160
and residents 168, 169
and urban development 81, 87
tower blocks, residential **39**, 128–32, 138

Tower Hill Botanical Garden,
    Worcester 93
towns 3
trade, international 26, 28–30
transport, public
    Chongqing 127, 132, 136,
        138
    Portland 97, 100–2, *104*,
        110
    Worcester 75, 96
trees 91
    Worcester 23, 73, 86–95,
        92, 96
    tree planting 90, 93, 94–5
Trump, Donald 120
Tuan, Yi-Fu 7

underdevelopment, Global
    South 135
Union Station, Worcester 74,
    75
United Kingdom 42, **151**
    see also City of London,
        United Kingdom;
        London, United
        Kingdom
United Nations 176
United States **81**, 176
    and racial issues **111–13**,
        181
    suburbs 37, **113**
    see also Oregon, United
        States; Portland, Oregon,
        United States; Worcester,
        Massachusetts, United
        States
United States Department of
    Agriculture (USDA) 86,
    87
universities 52, 68, 74, **167**
University of Massachusetts
    Medical School 74
Unum insurance company 75
urban areas, definition 3–5

urban development *see*
    development, urban
urban growth boundaries
    (UGBs) 97, 100, **101–3**,
    103
urban growth coalition 55–7
    Tehran 56, 68, 70
urban place imaginaries 18–21,
    89
urban policy 33–5, 87
urban processes 2, **15–17**,
    179–80
urban theory 39, **65**, **115–17**,
    156, 178
urbanism 65, 99, 176
    environmental 99–103,
        105–6, 183
    modernist 22, 27, 36–42,
        37–9, 44, 45, 46–7

Vancouver, Canada 98
Vanport, Portland, Oregon
    109–10
verticality, Chongqing 127–32,
    133
Virtually Asian video essay
    (Suparak) 135
Volksfront, white separatist
    group 116

wages, Tehran 54
water supplies, Tehran 58, 69
wealth **111**
    City of London 27–8, 32,
        42, 46–7, 177–8, 182
    Tehran 54, 60, 177–8
weather events 86
weirdness
    Chongqing 132–3, 136–8
    Portland 102, 105
West Bank, Jerusalem 153,
    157, 158, 169, 172
Western Wall, Jerusalem (Kotel)
    159, 163–4, *165*, 168

White Elephant (Baixiang) Street, Chongqing 128, 138
white people 113, 117, **119–21**
   Portland 98, 106–12, 114, 121
Willamette Valley, Oregon 99, 100, 103
William the Conqueror 26
Wirth, Louis 5, **65**
women 63–4, **65**, 66
women's clothing, Tehran 63, 66, 193n6, 193n7
Worcester, Massachusetts, United States 23, 72–96, 177
   baseball 80–2, 83–4
   collective identity 23, 78, 82, 94–5
   commuting 72, 74
   consumption 83, 95–6
   developers 74, 76–8, 83
   economic restructuring 23, 72–96, 183
   educational opportunities 74, 78
   entertainment 74, 83–4
   environmental issues 23, 73, 85–93, 183
      environmental policy 85–90, 93
      forests 73, 86, 88, 90–5
      longhorned beetles 85–8, 94
      trees 23, 73, 86–95, 92, 96
   government funding 76, 82
   household income 93, 94
   infrastructure/transport 75, 96, 183
   investment 73, 76–8, 80
   luxury housing 78, 82
   malls 74–5, 84
   population numbers 72, 73
   postcard campaign 80, 83–4
   real estate 73, 75
   redevelopment 76, 183
   residents 80, 83–4, 85, 87–93, 94–5, 96
      social class 82, 93
   tax revenues 75–6, 80, 95–6
   tax-increment financing/ district improvement financing 76, 82
   urban elites 80, 85
Worcester Tree Initiative (WTI) 90, 93
World War II, Portland, Oregon, United States 99, 109, 114
Wright. W. J. **67**, **119**

Yiftachel, Oren 156
Young, Iris Marion **15**
young people 78
   Portland 102, 112, 121
   Tehran 60, 68

Zhang, Amy 123–52, 184
Zhang, T. 57
Zukin, Sharon **67**